Encyclopaedia of Life Science

Encyclopaedia of Life Science

VOL. 3

B.A. Mannan

ANMOL PUBLICATIONS PVT. LTD.
NEW DELHI - 110 002 (INDIA)

ANMOL PUBLICATIONS PVT. LTD.
H.O.: 4374/4B, Ansari Road, Darya Ganj,
New Delhi-110 002 (India)
Ph.: 23278000, 23261597
B.O.: No. 1015, Ist Main Road, BSK IIIrd Stage
IIIrd Phase, IIIrd Block,
Bangalore - 560 085 (India)
Visit us at: www.anmolpublications.com

Encyclopaedia of Life Science

First Published, 2007

ISBN 978-81-261-3408-3 (Set)

PRINTED IN INDIA

Printed at Amit Enterprises, Delhi.

Contents

Preface

'Life Science' is a branch of science, under which we study about the living beings in detail. We study about the body structure; all the organs of the body and their functions. The body consists of several organs, which are made of Tissues, while, Tissues are made of cells. Plants are very useful for living organisms as they supply oxygen for the respiration and they are the food-source for the primary consumers. Plant survival and growth is critically influenced by abiotic factors including water, air and light. The effect of the local environment on plant growth also accounts for much 'of variation in adult form within a species. Plants have green pigments, which are quite useful in the process of photosynthesis. In the presence of light, plants make their food with the help of minerals, water and carbondioxide, in their green leaves, in the form of carbohydrates. The food prepared in green leaves is transported to different parts of the plant by phloem.

Similarly, animal life is a significant feature on earth. All the functions of the human beings and animals, including birds and insects, are performed in their cells. Cell is the structural unit of any living body. This was first discovered by scientist, Robert Hook. The cell has several parts, such as nucleus, centrosomes, mitochondria, plastids, cell-walls. The cell wall gives the mechanical support to the cell and cell-walls are accountable for the transportation of the material in and out of the cells.

Biologists agree that all living things arose through long periods of change shaped by physical and chemical processes. Approximately, there are two million species in the living world. In changing environment, a few species are eliminated during un-favourable conditions on the basis of the principle of the 'survival of the fittest,' which is also known as natural selection.

Science can strive to grasp nature's faculty, but science is socially embedded and all scientists record prevailing certainties. Darwin himself, in the closing lines of 'The Origin of Species', mentioned about Victorial social preference. Life's pathway certainly includes many features, predicable from laws of nature. But laws of nature do not tell us why we have crabs and snails at all, why insects rule the multicellular world, and why vertebrates rather than persistent algal mats exist as the most complex form of life on earth.

Bacteria too represent the great success story of life's pathway. They occupy a wider domain of environments and span a border range of biochemistries than any other group. One might conclude that complexification of life as a whole represents a pseudo-trend based on constraints at the wall, but still holds that evaluation continues, within a particular group.

This eminent work, titled; *Encyclopaedia of Life Science* would be able to make new waves in the discipline. Expectedly, these books would prove to be the best source of study in the area of Life Science.

Editor

1

Significance of Pheromones

Pheromones are chemical substances secreted by an animal externally with specific effects on the behaviour or physiology of another individual of the same species. The term pheromone was coined by *Karlson* and *Luscher* in 1959.

The immediate and invariable response produced by pheromones (from the Greek *pherein,* to carry; *hormone,* to excite) is comparable to the action produced by hormones. But unlike hormones, which work *within* an individual, these chemicals also known as ectohormones, work *between* the individuals. When these chemicals are used to communicate with other species, and the receiver benefits, they are called *Kairomones.* If it is the sender who benefits, the signal is an *Allomone.* These substances may be secreted rather generally by the skin or by specialised glands and, similarly, their detection by the recipient individual may be simply by ingestion or by specialised chemi-receptors such as those on the antennae of the silkworm moth.

Characteristics of Pheromones:

(1) Pheromones secreted by most of the animals are

proteins of low molecular weight or fatty acid derivatives.

(2) Pheromones are generally volatile substances and spread in the nature by vaporisation and diffusion.

(3) The maximum effect of pheromones depends on their concentration and vicinity to the animals effected.

(4) Pheromones having a long carbon chain serve as sex attractants whereas small carbon chained pheromones alert the animals against dangers.

(5) Sex attracting pheromones are generally released during the night and they can travel along the wind upto several kms.

Pheromones in Insects and their Application

E.F. Knipling and M. Viroza were the pioneers who identified and extracted pheromones. They also synthesised artificial pheromones and used them in insect control.

Classification of Pheromones: Pheromones are classified (i) according to the response of the recipient and (i) on the basis of their biological function.

According to Response of the Recipient

(i) *Releaser Pheromones:* These are typically odorous and act directly on the central nervous system of the recipient. Releaser pheromones produce an immediate change in the behaviour of the recipient, *e.g.*, the sexattractants of moths, and the trail pheromones and alarm substances released by ants.

(ii) *Primer Pheromones:* These act slowly and bring about a chain of physiological events in the recipient without any immediate change in behaviour. In a true sense, the body is "primed" for new biological activity, and control the number of reproductive members in social insects like termites, ants and bees.

On the basis of Biological Function

(i) *Trail Pheromones:* The social insects like ants and bees secrete a trail pheromone. The workers of ants such as *Solenopsis,* on their way back after finding a source of food, release a highly specific, volatile "trail pheromone" marking the trail in the form of a series of scent spots so that fellow ants find their way to the source of food. The pheromone is produced in Dufour's gland (poison gland), from Pavan's gland or from the hind gut. Accordingly, the secretion is released through the sting, the edge of the 6th abdominal sternite or the anus, as the case may be.

Worker bees, on finding a source of food, secrete a C_{10} alcohol called *geraniol* which leads other worker bees to the food.

(ii) *Alerting or Alarm Pheromones:* Many species of ants secrete "alarm pheromones" if the colony is disturbed. The workers are stimulated by a pheromone from the mandibular or anal glands which is released by any individual of the colony being disturbed. The pheromone transmits the alarm to the other ants in the area. As a result, they react by rushing about in a very agitated manner or assume an aggressive posture.

(iii) *Sex Pheromones:* Olfactory sex pheromones, also called "sex attractants" have now been established in more than thirty species of moths, butterflies, flies, beetles and cockroaches. In general males are attracted to females, but in the case of boll weevil, *Anthonomus grandis,* the males produce the attractant. In some other cases, aggregations occur in which both sexes are brought mutually together by these substances. For example, aphrodisiacs of many *Lepidoptera* and *Hymenoptera* are usually produced by we male after it has responded to the attractant of the female. Pheromones of the male then stimulate the copulatory behaviour of the female which is essential for successful mating.

The recipients of the sex pheromones respond in one of the two ways:

1. They may be stimulated to approach the releasing insect from some distance.

Or

2. They may be stimulated to perform some close range sex foreplay like courtship.

A few examples of sex pheromones are as follows:

1. *Bombykol:* Female silkworm moth *Bombyx mori* produces a pheromone called *bombykol* which attracts males from astonishing distances. It is detected by the male with the help of very fine alfactory sensory cells present on its antennae. As a result, the male starts flying upwind towards the female that is producing the pheromone. When male locates her, they copulate. *Bombykol* active at 10^{12} μ g, is an unsaturated C_{16} fatty alcohol with two double bonds:

$$CH_3 - (CH_2)_2 - CH = CH - CH = CH - (CH_2)_8 - CH_2OH$$

Bombykol

2. *Disparlure:* The female gypsy moth, *Porthetria dispar* secretes a sex pheromone called *disparlure.* Like bombykol, it, too is active to 10^{-12} μ g. Depending on wind speed, gypsy moth female can attract male from upto 4500 metres away.
3. *Antiqueen Pheromone:* In social insects such as honey bees, the queen secretes an antiqueen pheromone which acts on the corpora allata of the developing females and prevent their development into queens.

When the queen dies, the antiqueen pheromone is no longer present, and consequently one or more of the nymphs develop into queens. The same pheromone, however, serves, as a sex attractant to the drones during nuptial flight.

$$CH_3 . CO(CH_2)_5 \, CH = CHCOOH$$

Honey bee antiqueen substance

4. *Cabbage looper sex pheromone.*

$$CH_3(CH_2)_3\ CH = CH\ (CH_2)_6\ O — COCH_3$$

Sex pheromone of cabbage looper, *Trichoplusiani*

5. *Cotton pink boll worm sex pheromone.*

$$(CH_3CH_2CH_2)_2\ C = CH — (CH_2)_2 — CH = CH—(CH_2)_4\ OCOCH_3.$$

Sex pheromone of cotton pink boll worm,

Pectinophora gossypiella

Sex pheromones have been reported in still some other insects like peach stem borer, *sanninoida exitiosa* ; tobacco horn worm, *Protoparce sexia*; sugar cane stem borer, *Diatraea saccharialis*; cucumber beetle, *Biabrotica balteata* and coddling moth, *Carpocapsa domorella*.

(iv) Aggregating pheromones

(v) Area making (territoriality) pheromones. Some animals use pheromones to mark their territorial and home ranges.

(vi) Recruiting pheromones.

Pheromones in Mammals and their Application

Pheromones are known to play a role in the reproductive physiology of mammals. The earliest chemical studies on mammalian pheromones were carried out on the musk deer. *Muscopyridine* has been found to be the active principle of the odour emitted by the musk deer.

Pheromones in Mice: It has been found experimentally that when female mice are placed in groups of four or more per cage, they show a higher incidence of anoestrus and psuedo pregnancy than when caged individually. If one male mouse is kept in the cage, the odour emitted by the male reduces the frequency of reproductive abnormalities. Further, the smell of the urine of a male, obviously a pheromone, terminates the pregnancy in a newly impregnated female and it becomes sexually receptive again.

Pheromones in Monkeys: In monkeys, the sex drive of the male is greater when he is exposed to a female at the time of ovulation than when he is exposed to female at another time of her cycle. The "pheromone message' sent by the female to the male in this situation is olfactory. This pheromone is vaginal in origin.

Pheromones in Humans: Only little information is available with reference to human pheromones. Certain fatty acids exhibiting pheromonal properties are found in increased amounts in human vaginal secretions at about midcycle. Pleasant, musk-like odour of a human male pheromone, *14-hydroxytetradecanoic acid,* is perceived only by sexually mature women, particularly at the time of ovulation. Males and young girls are relatively insensitive to this substance. Similarly, *copulins* have been identified in human female vaginal secretions that have an arousing effect on the males.

2

Significance of Blood

Blood may be defined as a highly complex, seemingly homogeneous fluid composed of a large variety of dissolved and suspended inorganic and organic substances as well as three kinds of cells or formed elements. The presence of these cells accounts for the designation of blood as a tissue (or a *liquid tissue).*

In very small animals, oxygen penetrates the surface and diffuses to all parts of the organism. But in larger animals *(annelids, molluscs, arthropods, echinoderms* and *chordates)* a specialised transport system, the blood, becomes necessary. However, gas transport is not the only function of the blood. In higher animals, the blood has a number of functions stated below.

Functions of Blood

1. *Transport of Gases:* It transports oxygen from the lungs to the tissues and CO_2 from the tissues to the lungs.
2. *Transport of Nutrients:* It carries nutrients from the intestinal wall to various parts of the body, where they are used or stored.

3. *Transport of Waste Products:* Blood carries the excretory products from the cells or organs where they are formed to the excretory organs where they are eliminated.
4. *As a Communication System:* The blood also serves as a communication system within the body by transporting substances (vitamins, hormones, secretions, enzymes and other essential chemicals) that regulate the function of certain organs. For example, the hormone *secretin,* that is released from the wall of the small intestine into the blood, controls the secretion of pancreatic juice. It transports the secretions of endocrine glands to their places of activity.
5. *Defensive Action:* The blood acts as a great defensive mechanism against disease in two ways. *(a)* Its white blood cells engulf bacteria and foreign particles due to their phagocytic properties. *(b)* It develops *antibodies* which combat toxic agents.
6. *Coagulation:* The blood has the ability to coagulate and thus guards against *haemorrhage.*
7. *Functions of Plasma Proteins:* The plasma proteins of blood have various functions.
8. *Regulation of Body Temperature:* The water content of blood has three qualities which make it very suitable for this purpose, (a) *High-specific heat:* Due to this it can absorb a large amount of heat, thereby preventing sudden change of body temperature, (b) *High conductivity:* Thermal conductivity of water is higher than any other ordinary liquid. This helps in rapid and even distribution of heat, (c) *High latent heat of evaporation:* Latent heat of evaporation of water is very high and since water is constantly evaporated from skin and lungs, a large amount of heat is lost in this way.
9. *Regulation of Blood Pressure*: The blood has the ability to regulate its own pressure by changes in volume and viscosity (haematocrit value) of blood.

10. *Maintenance of Water Balance:* Among other factors blood also helps in maintaining an accurate balance between the water supply and loss.
11. *Maintenance of Acid-base Equilibrium:* By its efficient *buffering power (e.g.,* plasma proteins that are ionised, reduced, and oxidised to haemoglobin, etc.) and with the help of other organs it helps to maintain a constancy of acid-base content of all tissues including itself.
12. *Maintenance of Ion Balance:* The blood helps to maintain balance of ions between the cells and surrounding fluid.
13. *Functions of Mineral ions of Plasma:* The mineral ions of plasma function as nutrients and cellular control agents, and are essential regulators of the osmotic pressure of the blood.
14. *Maintenance of Own Integrity and Composition:* By regulation of its osmotic pressure, ion content, acid-base content, and by controlling the number of its cells, the blood has the unique ability to maintain its own integrity and composition. The uniformity in the pH of the blood of various vertebrates results from corresponding adjustments in the acid-base balance, achieved primarily through the amount of sodium.

Constituents of Blood

The blood is a highly complex, liquid tissue composed of loose cells *or formed elements* and *plasma.* By volume, the blood is 55 per cent plasma, and the remaining 45 per cent formed elements or corpuscles. The formed elements have a specific gravity of 1.09 as compared with 1.03 for the plasma, thus accounting for their separation by gravitation or centrifugation. The blood *serum* is the fluid formed when the blood clots.

Blood Plasma: This fluid contains two groups of substances. One comprises the compounds that fluctuate in concentration in accordance with the changing rates of their supply and removal. In this category are a number of foods in transit, urea and other waste products in transit, hormones and vitamins in transit, and several other classes of materials present only occasionally. The second group includes the substances that are normally maintained at constant concentrations. In this category are water, most of the mineral ions, plasma proteins, and certain nutrients and waste products in transit.

(a) *Water:* The main constituent of blood plasma is water (about 90 per cent by weight of the plasma) which has several functions. It maintains blood volume and blood pressure. The blood water functions as the *transport vehicle* for all other plasma components, and for respiratory gases, and blood cells.

(b) *Mineral Ions:* Dissolved inorganic materials carried in the plasma water comprise the *mineral ions.* These include mainly the positively charged Na^+, K^+, Ca^{++} and Mg^{++} ions and the negatively charged phosphate (H_2PO_4, HPO_4^{2-} -, PO_4^{3-} , chlorine (Cl^-), bicarbonate (HCO_3^-), and sulphate (SO_4^{2-}) ions. Over half of the total mineral content of plasma is sodium chloride. The ions are kept at constant concentration through mineral supply of food, turnover in tissue cells, excretion and other factors.

The mineral ions of the plasma function as nutrients and cellular control agents, and are essential regulators of the *osmotic pressure* of the blood. Another crucial regulating function of the plasma minerals is pH *control* and *pH* balance between blood and the tissue. Any small change of pH is readjusted to normal by the *buffering action* of the mineral ions. Some of the mineral ions exert other important effects. For example, calcium ions regulate the sensitivity of nerves and muscles. Calcium also influences the gel-sol states within cells, and it is a component of the cementing substances between cells. Since all cells are in contact with blood or lymph, the concentration of blood calcium evidently is of great significance.

(c) *Plasma Proteins:* Proteins make up about 7 to 8 per cent of the plasma and are classified into four general subdivisions: *serum albumins* (4.2 per cent), *serum globulins* (2.6 per cent), *fibrinogen* (0.3 per cent), and the remainder constituting 1 per cent or less of the plasma. These compounds endow blood with some of its most important regulating functions, including the function of chemical body defence. They maintain the osmotic pressure of the blood, and since most of them are ionised, they also play a role in maintaining the pH of the blood, like mineral ions.

Some of the plasma proteins *(e.g., fibrinogen)* are involved in the coagulation of blood. Many of the proteins are active enzymatically. One of the catalytic agents is a clotting factor *prothrombin.* Others are enzymes such as are found in tissue cells generally and even in the intestine. For example, *trypsin* is usually present in the plasma. The catalytic functions of such enzymes in plasma are still largely obscure. Similarly obscure are the specific roles of plasma proteins called the *albumins,* so named because of their chemical resemblance to egg white. The activity of certain enzymes in plasma is a useful index of some abnormal conditions. For example, *serum amylase* is elevated in acute pancreatitis. Plasma levels of *glutamate—oxaloacetate aminotransferase, lactate dehydrogenase,* and other enzymes have proved to be of some value as diagnostic aids for assessing cardiac damage. Elevation of the plasma levels of these and certain other enzymes, *e.g., aldolase,* also occurs in liver disease.

Proteins called *globulins* are very important and include α, β, and γ globulins. Several proteins are designated as α_1 or α_2 globulins. Some are of unknown function, such as α_1- *acid glycoprotein,* which has 42 per cent carbohydrate content, α*-foetoglobulin,* a glycoprotein is thought to serve functions in the foetus similar to those of albumin in the adult. *Retinol-binding protein,* which serves in retinal transport, forms an equimolar complex with *transthyretin,* which also functions as a thyroxine-binding protein. *Ceruloplasmin* has eight sites that bind either Cu^{+} or Cu^{2+}. *Haptoglobins* constitute about one-fourth of α_2- globulins and form complexes with haemoglobin. Such complexes are degraded by reticuloendothelial cells, and the iron is reutilised for haeme synthesis after degradation of globin and excretion of degraded haeme as bile pigment. Seven *protease inhibitors* have been found in human blood plasma, several of which are α-globulins.

The β-globulins of the plasma are a variety of proteins including the lipoproteins. Of them, *transferrin* constitutes about 3 per cent of the total plasma proteins. It regulates the concentration of free iron in plasma, thereby preventing iron accumulation in tissues and urinary loss of iron. *Haemopexin,* another β- globulin, binds haeme and prevents its urinary excretion, thereby retaining haeme iron for further use. *C-reactive protein* promotes phagocytosis and β_2 *microglobulin* is the smaller subunit of the HL-A histocompatibility antigen complex, which regulates rejection of transplanted tissues.

The γ-globulin fraction contains the immunoglobulins, which are the *antibodies* and serve to destroy antigens. Included in the group are the so-called cryoglobulins, immunoglobulins that precipitate on cooling serum. Cryoglobulins are often found during inflammatory illnesses, *e.g.*, rheumatoid arthritis, as well as in multiple myeloma.

(d) *Glucose and Other Simple Sugars:* The concentration of glucose in the plasma normally averages about 0.1 per cent. It is absorbed by the capillaries in the intestine, stored principally as glycogen in the liver and muscles, and is carried to all parts of the body by the blood to be used as a principal source of energy in cellular metabolism.

(e) *Other Constituents of Plasma:* The remaining 2 to 3 per cent of the plasma is made up of a variety of substances including amino acids, vitamins, lipids, hormones, urea, uric acid, and other metabolic waste products. In addition, plasma contains dissolved O_2

and CO_2 in transit to and from cells. All these substances fluctuate more or less widely in concentration in accordance with the changing rates of supply and removal.

Apart from the above constituents, blood plasma of certain animals also carries the blood pigments or respiratory pigments that transport the gases from the respiratory surface to the tissues and *vice versa.*

Blood Cells: The blood contains the following three kinds of formed elements or free cells which make up about 45 per cent of the blood.

(a) *Red Blood Cells:* The red blood cell or *erythrocyte* is by far the most prevalent of the formed elements. In man, each cubic millimetre of blood contains over 5 million red blood corpuscles. The enucleated mammalian erythrocytes lack DNA and the ability to synthesise proteins. Even with limited metabolic capability, these cells survive for weeks or months, depending on the species.

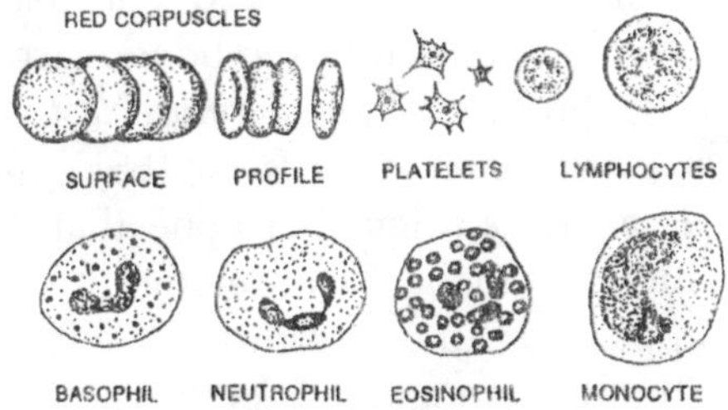

Formed elements of human blood.

Structure: In fishes, amphibians, reptiles, and birds the erythrocytes are oval, nucleated cells. Mammalian erythrocytes, on the other hand, are *non-nucleated* spherical discs, except in the camel group, whose red blood cells are oval.

Human erythrocytes average 7.5 μ in diameter, 2.4 μ in thickness at the edge, but only 1 μ thick in the centre. They are acidophilic and therefore, stained with eosin. Under electron microscope, mature erythrocytes lack an endoplasmic reticulum, Golgi apparatus, ribonucleoprotein and mitochondria. The human erythrocyte has a highly oriented structure of lipids, proteins and haemoglobin consisting of approximately 64 per cent water, 28 per cent haemoglobin (approximately 100 million haemoglobin molecules), 7 per cent lipids or fatty materials, and the remaining 3 per cent sugars, salts, enzymes and other proteins. Haemoglobin of the erythrocytes is responsible for the red colour of the blood.

Functions: Haemoglobin of the red blood corpuscles transports oxygen and enables our blood to carry approximately sixty times the amount of O_2 than an equal volume of water or plasma could transport in solution. Haemoglobin also functions as a carrier of a small quantity of CO_2, conveying it as a waste product from the tissues to the lungs for excretion.

Life History of the Red Blood Cell: The average life of a red blood cell is about 120 days. Under normal conditions they are continuously destroyed and new ones produced at the same rate of about 140 million per minute. In the embryo they are formed in the liver and spleen, and after birth, in the red bone marrow. The formation of red corpuscles is called *haemopoiesis* or *erythropoiesis* and a red bone-marrow cell which gives rise to an erythrocyte is called *erythroblast*. The substance called the *erthropoiesis stimulating factor* (ESF), also known as *erythropoietin,* is a glycoprotein hormone that stimulates the red cell production. Prior to entering the bloodstream, erythrocytes appear as undifferentiated cells, each with a nucleus but without haemoglobin. As these cells mature, they undergo a series of changes which include disappearance of the nucleus (probably by extrusion or dissolution) and haemoglobin formation. Upon reaching maturity, they enter the bloodstream from the special connective tissue network of the red bone marrow in which they develop.

Control of Red Corpuscles Number: The major function of the red corpuscles is the transporting of respiratory gases. In the adult, red corpuscles are manufactured in the red marrow present at the ends of the long bones (*e.g.*, ribs, arms, legs). The liver and the spleen are the production sites in the embryo before the bones mature. After the skeleton is fully formed, the blood cell production is initiated in it, the spleen functioning principally a blood storing organ. It may contract like a sponge and squeeze reserve blood into circulation. The liver becomes the organ where the red corpuscles are destroyed.

Erythropoiesis is controlled by a feedback mechanism which depends upon the oxygen state of the blood. The reduced oxygen state of the blood affects the kidneys and the liver. As a result, the liver releases more globulin and the kidneys produce more *renal erythropoietic factor* (REF). The globulin and REF, interact in the blood to from ESF, which then stimulates greater red cell production with more haemoglobin, thus a greater quantity of oxygen can be carried which has a negative influence on the production of globulin and REF. At high altitudes, the low oxygen pressure stimulates the sequence just outlined, resulting in an elevated red cell count, a condition termed *polycythemia*. A persistently high oxygen concentration in the blood has the opposite effect on the liver and the bone marrow. Through this mechanism the number of red corpuscles is adjusted to oxygen requirements. Since such requirements and the composition of the atmosphere and hence the amount of oxygen normally remain fairly constant, the number of red corpuscles is also constant.

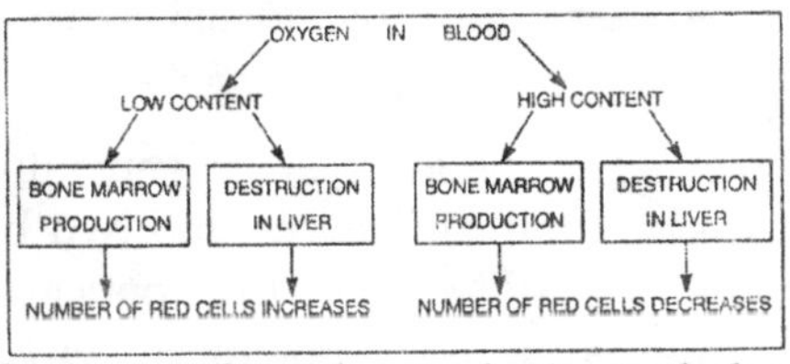

Control of red corpuscle number. Production in the bone marrow and destruction in the liver are geared to each other by the oxygen content of the blood.

When the number of red blood cells or their haemoglobin content or both are reduced significantly, the result is *anaemia.* This may be brought about by excessive corpuscle destruction in an abnormal liver, or by extensive blood loss through wounding. In the latter case, production in bone marrow soon makes up the loss. Anaemia may also arise as a result of inadequate production of corpuscles as, for example, when marrow cells are defective, or when the iron content of the diet is low.

However, the destruction of red corpuscles in the liver is random and non-selective, *i.e.*, it affects young as well as old corpuscles. The iron of disintegrated haemoglobin becomes generally available in metabolism and may be used, for example, in the synthesis of cytochrome or of new haemoglobin. The tetrapyrrole fraction of destroyed haemoglobin is largely excreted in bile and urine.

White Blood Cells and their Functions

White blood cells or *leucocytes* (leukocytes) are metabolically among the most active of the formed elements. About 7000 to 10,000 of these cells are found in each cubic millimetre of human blood. They are non-pigmented cells, *i.e.*, devoid of haemoglobin, contain a nucleus, and the majority are larger than erythrocytes, measuring from 8-15 μ in diameter depending on the type.

The term *leukopenia* describes a condition characterised by a low white cell count, which is usually associated with the viral diseases such as mumps, measles, chickenpox and poliomyelitis.

Structure and Types: Leucocytes are generally classified on the basis of the structural differences into two broad groups : (i) leucocytes with a single nucleus (mononuclear) and a clear cytoplasm (agranulocytes), and (ii) polymorphonuclear leucocytes or granulocytes.

The first group consists of two types of cells: the *lymphocytes* and *monocytes*, both arising from the lymphatic tissue. White blood cells of the second group have their origin in the red bone marrow, contain cytoplasmic granules (hence the name granulocytes) and display a nucleus with two or more lobes. They are subdivided into three types (neutrophils, eosinophils, basophils, according to the staining reactions of their cytoplasmic granules.

(a) *Lymphocytes:* They range in diameter from 8 to 12 μ and constitute about 25 per cent of the total number of leucocytes. They are small round cells with a more or less centrally placed nucleus. In a *small lymphocyte,* that is slightly larger than a red cell, nucleus is relatively large and occupies the major part of the cell. In a *large lymphocyte,* the cytoplasm is proportionally more and forms a wider zone around the nucleus.

(b) *Monocytes: A* monocyte is a large cell with a kidney or horse-shoe shaped nucleus. They range in size from 10-15 μ and are quite mobile. They represent 5 to 7 per cent of the white blood cells.

(c) *Polymorphs or neutrophils:* A neutrophil has an irregular lobulate nucleus having 2 to 7 lobes. Their name is derived from the intense stain that the neutral dyes confer on their cytoplasmic granules (granules stain pale-violet with Leishmann's stain). Neutrophils are approximately 12-15 μ in diameter, circular in outline, and are the most active of all the white blood cells in terms of motility and ingestion of foreign particles. They constitute about 60 to 70 per cent of the total number of leucocytes.

(d) *Eosinophils:* They are stained by the acid dye, eosin, and constitute 2 to 4 per cent of the total white blood cells. They measure about 12-15 μ in diameter and contain a bilobed nucleus.

(e) *Basophils:* They are circular in outline and measure about 12-15 μ in diameter. Typically their nuclei contain 2 to 3 lobes, are centrally placed, and are stained relatively faintly. The cytoplasm is moderately abundant and cytoplasmic granules dominate the appearance of basophilic leucocytes. Unlike eosinophilic and neutrophilic granules, basophilic granules overlie and obscure the nucleus.

Leukocytes are also classified as the *lymphocytes* and the *phagocytes*. The latter can be divided into two categories, the *polymorphonuclear leucocytes* and the *monocyte-macrophages*.

Functions of Leucocytes: The primary function of the leucocytes is to protect the body against many types of bacterial infections. Polymorpho-nuclear leucocytes are the first cells to arrive at sites of inflammation and are an important component of the host defence mechanism against bacterial infections. Macrophages function to scavenge cellular debris generated by polymorphonuclear leucocytes.

They also destroy those antigens that the polymorphonuclear leucocytes are unable to digest. In addition, macrophages perform the critical immunologic role of processing antigens into a form that allows lymphocytes to recognise them. The ability of leucocytes to destroy microorganisms depends in part on their motility and their capacity to ingest foreign particles. In this respect *neutrophils* and *monocytes* are the most active. By forming pseudopods or streaming extensions of protoplasm, they engulf and destroy foreign particles, including bacteria. This ingestion process is called *phagocytosis*.

Lymphocytes too serve in body defence, but they also contribute to scar-tissue formation after internal or external injury and in this way they facilitate wound healing. When lymphocytes are stimulated by their specific antigens, they produce a variety of soluble products that function as mediators and modulators of inflammatory process. These diverse substances are called *lymphokines.* All of the lymphokines studied so far appear to be proteins. Numerous lymphocytes are normally present in the lymph nodes, where they act as lymph-purifying agents. For example, microscopic particles of dust, smoke, and other materials present in the atmosphere frequently get into the lungs and may become embedded in the lung tissue. Lymph then carries such particles to the lymph nodes where the lymphocytes engulf the particles and retain them permanently. However, lymphocytes are responsible at least in part for the manufacture of both the beta-and gamma globulins of the plasma, and therefore for the immunity substances *(antibodies)*.

Eosinophils and the *basophils* are not very mobile and show only a slight tendency to engulf foreign particles. Their role is not well established. However, they have some relationship to disease resistance since their concentrations in the bloodstream are specifically increased during certain infections. This phenomenon is very helpful in the diagnosis of certain diseases. For example, the eosinophil count is significantly increased during infection by hookworm, as well as in allergic states such as asthma.

The number of the white cells increases temporarily during infectious diseases, as would be expected in view of the phagocytic activity of the cells. An abnormally hight white cell count is called *leukocytosis*. An uncontrolled and abnormally rapid formation of white blood cells leads to the cancer of bone marrow or *leukemia*. This condition, which is eventually fatal, is accompanied by anaemia because the bone marrow cells that normally give rise to red blood cells are displaced by leukemia cells.

Blood Platelets and their Functions

Blood platelets or *thrombocytes* are small, oval fragments of specialised bone marrow cells, measuring from 2 to 4 μ in diameter and generally lacking nuclei. About 250,000 of these cells are found in each cubic millimetre of human blood. Blood platelets contain centrally placed blue-violet granular material, the *granulomere*, which may give a false impression of a nucleus. The periphery of platelets is clear of granules, stains light blue, and is called the *hyalomere*. Sometimes thread-like processes extend from blood platelets.

Function of Platelets: The metabolic activity of blood platelets has recently been found to be equal to, if not greater than, that of the white blood cells. Blood platelets play an essential role in the coagulation of blood. The platelets have the capacity to absorb many of the factors required for blood clotting and carry them to the site of blood coagulation. Platelets are necessary for the intrinsic clotting process. They release a phospholipid, the 'platelet factors 3,' which participates in the conversion of prothrombin to thrombin by blood clotting factors X (10th) and V (5th). Platelets may also function to plug small defects in a blood vessel. Platelet *serotonin,* a vasoconstricting agent, may induce constriction of small blood vessels at the site of injury and may function to reduce blood loss.

Platelets are required for *clot retraction.* It is likely that the ATP and actinomyosin-like compounds in platelets are required for this phenomenon. *Tocatins* (1936) has described clumps of platelets adhering to fibrin strands. As these clumps are formed, fibrin becomes twisted and bent and clot retraction takes place. Knots of platelets are found at the intersections of fibrin strands and may strengthen the clot. Intimate association of platelets and fibrin has been confirmed by electron microscopy.

Smith et al (1983) have summarised the roles of platelets in coagulation as follows:

(i) They form the platelet plug in injured vessels by adhering to exposed subendothelial cell structures.

(ii) They are the site of activation of some clotting factors.

(iii) They provide the surface upon which certain clotting factors bind.

(iv) They are a source of certain clotting factors, *e.g.*, factor XIII and phospholipids.

Main Constituents and Functions of Blood plasma.

S. No.	*Component*	*Function*
1.	Water	Maintains blood volume and pressure, forms lymph, water supply of cells, provides vehicle for other constituents.
2.	Mineral ions	Maintain osmotic balance, pH balance,

contd....

S. No.	*Component*	*Function*
		buffer capacity, varied effects on tissue cells.
3.	Plasma proteins fibrinogen prothrombin albumins, enzymes globulins	All maintain osmotic and pH balance Participates in blood clotting Participates in blood clotting Functions obscure, basis of blood types, act as antibodies.
4.	Glucose and other metabolites	In transit to and from cells.
5.	O_2, Co_2, urea, various foods, hormones, vitamins, etc.	In transit to and from cells,

Cellular Constituents of Human Blood and their Functions

Component	*Origin*	*Size (diametre)*	*Appr. Number per cubic millimetre*	*Functions*
Erythrocytes	Spleen, liver (embryo) red, bone marrow (adults)	7 - 8 μ	5 million	Transport of O_2 and CO_2
Leucocytes	Red bone marrow	8 -15 μ	8000	Engulf bacteria, foreign bodies, aid in wound healing
Lymphocytes	Lymph nodes	8 -12 μ	20 - 25 per cent of total white cells.	—do—
Thrombocytes (platelets)	Red bone marrow, lungs	2-4 μ	250,000 to 600,000	Release thromboplastin, aid in blood clotting

A deficiency of platelets is called *thrombocytopenia.* It can be caused by many toxic agents, including certain drugs, ionising radiation, and some types of infections. In *thrombocytopenia purpura,* platelet levels in the blood are quite low, large extravasations of blood, or *purpura,* take place. Such blood disorders as leukemia may also be accompanied by thrombocytopenia.

Physiology of Blood Coagulation: When blood is shed it loses its fluidity in a few minutes and sets into a semi-solid jelly. This phenomenon is called *coagulation* or *clotting.* On keeping further, the clot retracts to a smaller volume and presses out a clear straw-coloured fluid, called the *serum.* Serum will not clot any more. The blood coagulation is a very complicated process and is under sensitive control. The coagulation is the property of plasma alone. The red blood corpuscles do not take part in it. They only become caught up in the meshes of the clot and are thereby removed. It is due to this fact that the clot has a red colour, and the serum is a clear non-cellular fluid. Blood platelets take part in coagulation of blood. They become sticky and rapidly adhere to the damaged blood vessels by binding to endothelial connective tissue.

The platelets also adhere to one another and finally form a plug, which is sufficient to stop bleeding if damage is small. The cessation of bleeding is called *haemostasis.* As the platelets aggregate, they release vasoactive amines, *e.g.,* serotonin and epinephrine, and prostaglandin metabolites, *e.g.,* thromboxane A_2, which stimulate vasoconstriction. Blood coagulation is then initiated around the platelets and the damaged tissues, leading to the formation of a *thrombus,* or blood clot, the major chemical defence against blood loss.

The following theories have been proposed to explain the mechanism of clotting of blood.

Best and Taylor's Theory: According to this theory, four substances are necessary for coagulation. These are: *prothrombin, calcium ions, thromboplastin* and *fibrinogen.* Prothrombin is produced in the liver and is present in small quantities in the blood plasma also. Thromboplastin is a fatty compound, found in the tissues. Calcium occurs as an inorganic constituent in the plasma. The optimum concentration required is 5 mg per 100 ml and the normal blood contains about this level. Fibrinogen is a soluble protein, found dissolved in the plasma.

When blood is shed, thromboplastin (thrombokinase) is liberated from the damaged tissue and from the disintegrated blood platelets. It acts on prothrombin in the presence of calcium, converting it to active *thrombin*. Now thrombin reacts with fibrinogen, and converts it into fibrin, which is insoluble. The latter constitutes the blood clot or *thrombus* in which blood corpuscles are trapped.

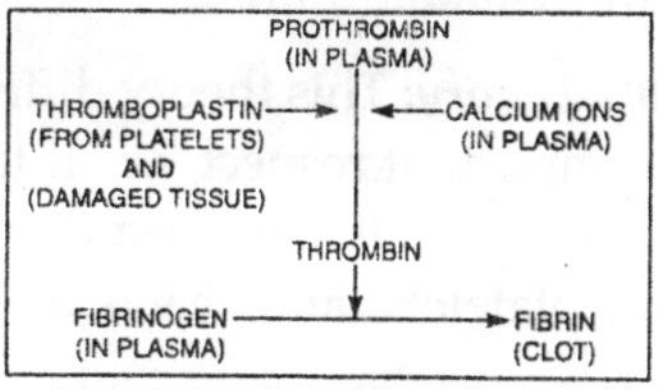

The main events of blood clotting (coagulation).

Blood within the blood vessels does not clot due to the lack of free thromboplastin, which is necessary to convert prothrombin into thrombin. If by chance thrombin is formed inside the blood vessels, it is immediately neutralised by another substance called *anti-thrombin*, which is normally present in the blood in very small quantities.

Howell's Theory: It states that clotting mechanism of mammalian blood takes place in two distinct stages:

(a) formation of thrombin from prothrombin.

(b) action of thrombin on fibrinogen.

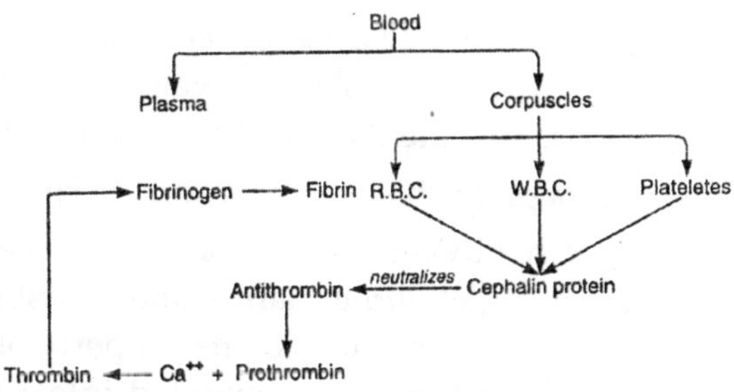

Mechanism of blood clotting (according to Howell).

However, it explains the formation of thrombin from prothrombin by the action of calcium alone.

The blood does not clot in the blood vessels because of the presence of *anti-thrombin* which prevents the activation of prothrombin. On the injury of blood vessels, tissue cells produce a substance known as *cephalin protein* which neutralises the anti-thrombin and thus allows calcium to react with prothrombin.

As a result, thrombin is formed, which reacts with firbrinogen to form insoluble fibrin.

Fuld and Spiro's Theory: This theory differs from Howell's theory only in one important respect, *i.e.*, it takes into account the presence of the enzyme *thrombokinase* which is produced especially by blood platelets, and reacts as follows:

(a) Thrombokinase + calcium ions + prothrombin → thrombin

(b) Thrombin + fibrinogen → Fibrin (clot)

Blood Rh-factors

In 1954, an International Commission on Blood Coagulation suggested a system of nomenclature of blood coagulation factors, which are designated by Roman numerals according to the recommendations of the commission.

These factors are first dealt with in numerical order, which indicates the historical sequence of their discovery.

(i) *Factor I or Fibrinogen:* Plasma concentration of fibrinogen is 300 mg/dl and its molecular weight is 340,000. The molecule has two identical subunits, each consisting of three polypeptide chains, α, β, and γ. The chains are bound together by disulphide bonds at each end of the molecule. It is globulin in nature and is distinguished from other plasma proteins by its property of clotting, during which *firbinogen* is converted into *fibrin.*

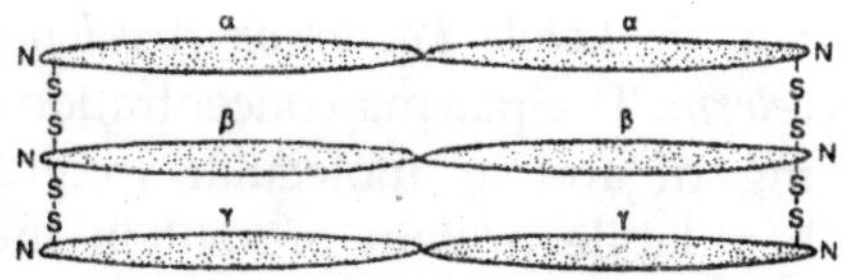

The structure of fibrinogen molecule consisting of two identical subunits, each having three polypeptide chains (α, β, and γ) which are bound together by disulphide bonds at each end of the molecule. The N-terminal amino acids are indicated by N.

(ii) *Factor II or Prothrombin:* Plasma concentration of prothrombin is 15 mg/dl and its molecular weight is 72500. It is protein in nature and is manufactured in the liver. However, vitamin K is essential for the formation of prothrombin. During clotting *prothrombin* is converted into *thrombin* by means of *prothrombin activator* which is also called thromboplastin.

(iii) *Factor III or Thromboplastin or Tissue Factor:* It is a protein having a molecular weight 43000. It is derived from two sources: intrinsic in the plasma, and *extrinsic* in the tissues. *Intrinsic thromboplastin* is formed in the plasma due to interaction between different plasma factors, *e.g.*, factor XII (Hageman factor), factor XI (PTA), factor IX (Christmas factor), factor VIII (antihaemophilic factor), calcium ions, factor V and factor X. *Extrinsic* or *tissue thromboplastin* is formed from different tissues, e.g., extracts of lungs, placenta, brain, etc., as a result of injury. So far it was known that prothrombin was converted into thrombin with the help of calcium ions and thromboplastin liberated from the damaged tissues. But recently it has been found that various plasma factors, *e.g.*, factor VII or *proconvertin*, are required for such conversion and the process is called *extrinsic thromboplastin formation.*

(iv) *Factor IV or Calcium:* The plasma concentration of Ca^{2+} ions is 0.5 mg/dl. Ionic calcium acts as a *cofactor* in coagulation. It is essential for the formation of both intrinsic and extrinsic thromboplastins and also in the conversion of prothrombin into thrombin.

(v) *Factor V or Labile Factor or Accelerator Globulin or Proaccelerin:* The plasma concentration of proaccelerin is 1 mg/dl and its molecular weight is 330,000. It is a heat-labile protein present in the plasma. This factor is necessary for the complete conversion of prothrombin into thrombin by the extrinsic or intrinsic thromboplastin.

(vi) *Factor VI or Accelerin or Va:* This factor is a hypothetical activation product of proaccelerin (factor V).

(vii) *Factor VII or Stable Factor or Proconvertin:* This factor is present in the plasma and is not used up during clotting. It is a protein associated with prothrombin. The plasma concentration of proconvertin is 0.05 mg/ dl and its molecular weight is 45,500. It accelerates the formation of tissue thromboplastin, being activated by extract released from the damaged tissue. During blood clotting proconvertin is changed into convertin.

(viii) *Factor VIII or anti-haemophilic Factor (AHF) or anti-haemophilic Globulin (AHG) or Platelet Co-factor I:* The plasma concentration of anti-haemophilic/von Willebrand factor is 1.6 mg/dl and its molecular weight is l-2×10^6, having 15 per cent carbohydrate content. This factor helps in the formation of intrinsic thromboplastin and intrinsic prothrombin conversion. It is present in the plasma and disappears when the blood clots. It is a protein closely associated to fibrinogen. This factor is anti-haemophilic.

In *Haemophilia* (bleeder's disease) the defect is not in the platelets but it is due to the absence of this factor which helps in the breakdown of platelets and liberation of platelet cofactor I or thromboplastinogen. The deficiency of AHG resulting in classical haemophilia in the male is transmitted as a sex-linked recessive character. The body fails to synthesise this essential globulin due to the absence of the specific enzyme which is controlled by the mutant gene.

(ix) *Factor IX or Christmas Factor or Plasma Thromboplastin Component (PTC) or Platelet Co-factor II or anti-haemophilic Factor B:* The plasma concentration of Christmas factor is 0.3 mg/dl and its molecular weight is 57,000. It has 26 per cent carbohydrate content. This factor is necessary for intrinsic thromboplastin formation. The absence of this factor leads to a disease known as *Haemophilia C.* This type of disease was first seen in a patient named *Christmas,* hence the name Christmas factor. This factor is not used up during clotting.

(x) *Factor X or Stuart-Prower Factor:* The international nomenclature has been given to this factor in 1959. The plasma concentration of this factor is 1.5 mg/dl and its molecular weight is 59,000. It has 15 per cent carbohydrate content. It is a circulating plasma zymogen, which on activation by factor IXa serves to activate prothrombin to thrombin.

(xi) *Factor XI or Plasma Thromboplastin Antecedent (PTA) or anti-haemophilic Factor C:* The plasma concentration of factor XI is 0.6 mg/dl and its molecular weight is 160,000. It has 12 per cent carbohydrate content. This is activated by the active Hageman factor (XIIa) and ultimately leads to the formation of thrombin. Deficiency of this factor leads to haemophiloid D and is transmitted as a sex-linked dominant in both the sexes.

(xii) *Factor XII or Hageman or Surface Factor:* The plasma concentration of factor XII is 2.9 mg/dl and its molecular weight is 76, 000. It has 15 per cent carbohydrate content. This is a protein in nature. Inactive form is activated on surface contact. This is also known as the clot-promoting factor or glass factor.

(xiii) *Factor XIII or Fibrin-stabilising or Laki-Lorand Factor (LLF) or Fibrinase:* The active form along with calcium ions converts soft fibrin clot to a solid, fibrous state. The plasma concentration of factor XIII is 2.0 mg/dl and its molecular weight is 320,000. It has 2.5 per cent carbohydrate content.

Other Factors: High-molecular-weight kininogen (HMW-K) or Fitzgerald factor:

This factor catalyses the initial reaction in which inactive factor XII is converted to active factor XII (XII a).

Pre-Kallikrein (Pre-K) or Fletcher Factor: Pre-Kallikrein or inactive plasma kallikrein is activated by pre-kallikrein activators, which are proteolytic fragments of the active form of clotting factor XII (Hageman factor).

Kallikrein: Plasma kallikrein also catalyses the activation of factor XII.

Platelet Phospholipid (PL): Phospholipids from aggregated platelets (PL) and Ca^{2+} are necessary for full activation of factor X. This also catalyses the conversion of prothrombin to thrombin.

Protein C: Although not listed as clotting factors, protein C is often referred to as factor XIV. Its molecular weight is 62,000. It is a circulating zymogen of protease which on activation by thrombin specifically inactivates factors V and VIII by proteolysis. Congenital absence of protein C leads to uncontrolled intravascular clotting and, in general, death in infancy.

Some features of human blood coagulation factors are summarised in Table:

The Human Blood Coagulation Factors and Some of their Properties.

Factor No.	*Name*	*Plasma concentration mg/dl*	*Molecular weight*	*Carbohydrate content %*
I	Fibrinogen	300	340,000	4
II	Prothrombin	15	72,500	8
III	Thromboplastin	...	43,400	...
IV	Calcium ions	0.5	....	...

Contd...

Factor No.	*Name*	*Plasma concentration mg/dl*	*Molecular weight*	*Carbohydrate content %*
V	Proaccelerin	1	330,000	?
VI	Accelerin	....	......	...
VII	Proconvertin	0.05	45,500	...
VIII	Antihaemophilic factor A	1.6	1-2 × 10^6	15
IX	Christmas factor	0.3	57,000	26
X	Stuart factor	1.5	59,000	15
XI	Plasma thromboplastin antecedent	0.6	160,000	12
XII	Hageman factor	2.9	76,000	15
XIII	Fibrin-stabilising factor	2.0	320,000	2.5

Role of Thrombin: It is a homogeneous glycoprotein. It autocatalyses its own formation by converting plasma Ac-globulin into active serum Ac-globulin and by labilising platelets to accelerate thromboplastin generation and liberates vasoconstrictors. By acting as a proteinase, it splits only four peptide bonds in the conversion of fibrinogen to fibrin and these are all bonds between arginine and glycine residues.

Role of Phospholipid: Phospholipid *kephalin* (cephalin) helps in the formation of *prothrombinase.* In the intrinsic system it is in the platelet factor 3, a phospholipid which participates in the conversion of prothrombin to thrombin by factors X and V, and in the extrinsic in tissue thromboplastin.

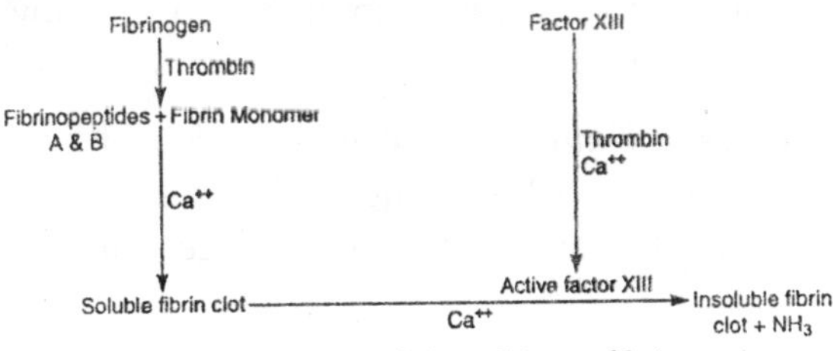

Showing the action of thrombin on fibrinogen.

Role of Vitamin K: Normal production in the liver of

prothrombin as well as factors VII, IX and X requires adequate dietary intake of vitamin K (*koagulation* vitamin). Vitamin K is required to form the binding sites for Ca^{2+} in prothrombin as well as in factors VII, IX and X.

The Ca^{2+} binding sites are formed by clusters of y⁻carboxyglutamate residues in the amino-terminal regions of each factor.

The anti-metabolites of vitamin K, such as *dicumarol,* inhibit the synthesis of these factors. Vitamin K is an essential cofactor for a *carboxylase* located on the lumenal side of the rough endoplasmic reticulum that carboxylates glutamyl residues in newly synthesised vitamin K-dependent clotting factors.

Prothrombin Activator: This is formed in two ways: (i) as a result of tissue (extrinsic system) and (ii) by activation of an intrinsic system consisting entirely of the blood constituents. In both the cases the key reaction is the conversion of factor X to its active form (Xa). The latter then interacts with phospholipid and factor V to form prothrombin activator. In the extrinsic system the reactions are as given below:

$$\text{Factor X} \xrightarrow{\text{Tissue factor, Factor VII, } Ca^{2+}} X_a$$

$$X_a \xrightarrow{\text{Phospholipid, FactorV}} \text{Prothrombin activator}$$

(Prothrombinase)

However, in the intrinsic system the reactions leading to the formation of Xa are more complicated as shown in the following hypothesis.

Hypothesis of Davie and Ratnoff: *Davie* and *Ratnoff* (1965) have proposed the so-called 'waterfall' sequence hypothesis to explain the succession of events taking place in the coagulation of blood.

The sequence of events leading to coagulation depends upon whether *thromboplasin* is present or not. It is the active substance in aqueous extracts of all tissues. Thus, a wound that results in bleeding also disrupts tissue causing the liberation of thromboplastin. Because a substance is required that is not a normal constituent of blood, this sequence is referred to as the *extrinsic system of coagulation.* On the other hand when all the substances involved are normally found in the blood, this sequence is referred to as the *intrinsic system of coagulation.* In both the systems each coagulation factor exists in the plasma in an inactive (proenzyme) form and is activated sequentially until thrombin is formed, which then converts fibrinogen into fibrin.

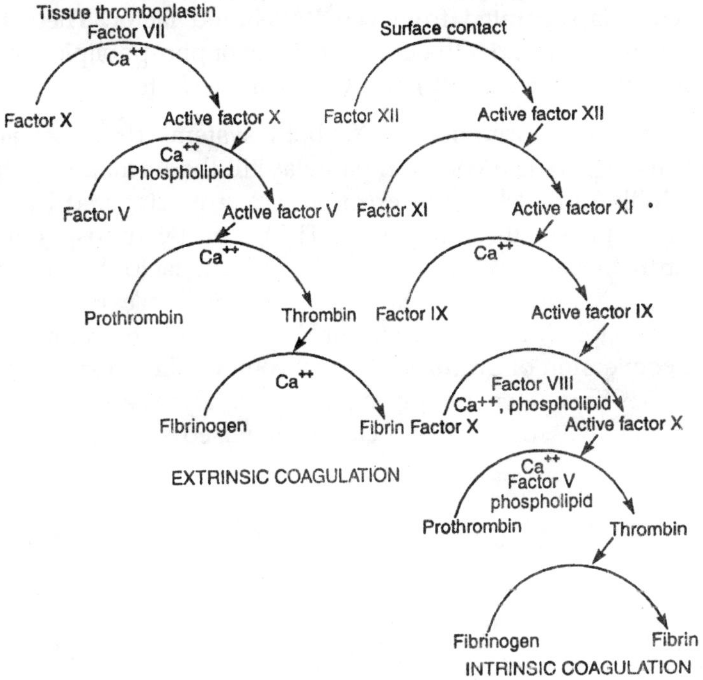

The "waterfall' schemes of extrinsic and intrinsic blood coagulation according to Davie and Ratnoff.

Enzyme Cascade Assumption

According to this hypothesis, suggested by *Macfarlane* (1967), the clotting mechanism consists of a large number of enzymatic reactions.

The term "cascade" expresses the relationship whereby an enzyme acts on a precursor, converting it to an enzyme that then acts on another precursor, converting it to an active enzyme and so on. In the *intrinsic system,* the initial reaction is conversion of inactive factor XII to active factor XII (XIIa). This is catalysed by high-molecular-weight kininogen (HMW-K) and kallikrein (Ka). This activation usually occurs when blood is exposed to the collagen fibres underlying the endothelium in the blood vessels. Active factor XII (XIIa) then activities factor XI, and active factor XI (XIa) activates factor IX. Activated factor IX (IXa) forms a complex with active factor VIII, which is activated when it is separated from von Willebrand factor. Then, the complex of IXa and VIIIa in the presence of phospholipids from aggregated platelets (PL) and Ca^{2+} activate factor X.

After the formation of Xa both systems (intrinsic and extrinsic) pursue a common pathway till the formation of the insoluble fibrin clot. The *extrinsic system* is triggered by the release of tissue thromboplastin (TPL), a protein phospholipid mixture that activates factor VII. The TPL and factor VII activate factors IX and X. Thereafter, in the presence of platelet phospholipid (PL), Ca^{2+} and factor V, activated factor X catalyses the conversion of prothrombin to thrombin. Finally, thrombin converts fibrinogen into fibrin and also catalyses the activation of factor XIII so that fibrin clot is stabilised.

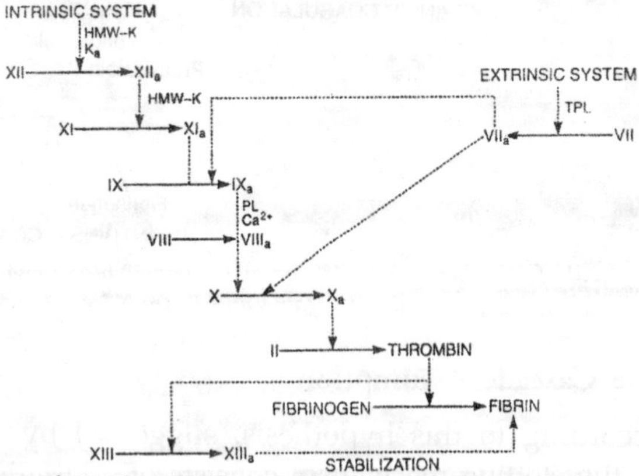

The scheme of reactions leading to blood coagulation by activation of intrinsic and extrinsic systems.

Another cascade, in which the enzyme *tissue plasminogen activator* converts the inactive precursor *plasminogen* to fibrin specific proteolytic enzyme *plasmin*, brings about the break down of the insoluble fibrin clot. This is called clot lysis pathway.

One very important aspect of the enzyme cascade hypothesis is that it is an amplifying system so that minute amounts of earlier factors can lead to final rapid conversion of large amounts of fibrinogen to fibrin. For example, *Esnouf* and *Macfarlane* (1968) quote that the clotting of 1 ml of blood involves 3 *mg* of fibrinogen, 100 μg of factor II, 10 μg of factor X, 0.1 μg of factor VIII, and probably less of the earlier factors.

However, two comments on the enzyme cascade hypothesis must be made. Firstly, it is an over-simplification of an extremely complicated process. Because it has not yet been shown that all the reactions involved are enzyme-substrate in type. Secondly, there are some additional reactions. For example, thrombin exerts positive feedback effects on the factors concerned in the formation of prothrombin activator. It activates factor VIII, perhaps by releasing it from its attachment to fibrinogen. It also increases the activity of factor V and XIII, and it also promotes aggregation of platelets, thus increasing the amounts of available phospholipid. However, after activating these mechanisms, thrombin soon inactivates them. Thus, thrombin intensifies the reactions promoting the local clotting, but only for a short time, i.e., until a firm clot has been formed. The spread of clot formation from the site of injury does not normally occur, partly because of the reactions just mentioned which cut it short, and partly because blood contains inhibitors of probably all the factors known to be concerned with the mechanism of clotting.

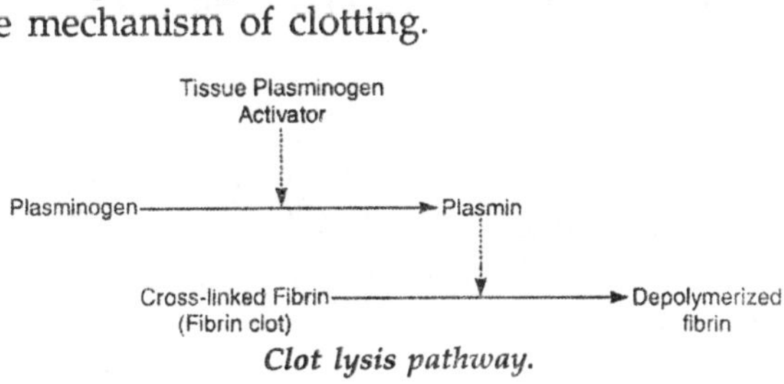

Clot lysis pathway.

According to Loewy et al (1991), clotting is an example of how post-translational modification—the covalent modification of proteins after the translation process—is used to regulate assembly processes.

Intravascular Clotting or Thrombosis: Thrombus is a clot formed in the living heart or inside the blood vessels from constituents of the blood. This is formed due to slowing of circulation and damage of the vascular endothelium. Masses of platelets are deposited on the damaged endothelium. Filaments of fibrin also form a network in this region.

The platelets liberate thromboplastin. The fibrin, entangled in the lamellae of platelets, forms the thrombus or clot. Intravascular thrombosis sometimes occurs in coronary and cerebral vessels which are called *coronary thrombosis* and *cerebral thrombosis* respectively.

Fibrinolysis: Clotted blood remains intact for several weeks if kept sterile. But if it is not kept sterile, the clot breaks up. This fibrin breakdown in the clot is known *as fibrinolysis* and is brought about by a proteolytic enzyme in the plasma known as *plasmin* or *fibrinolysin.*

The precursor of the enzyme, *plasminogen* (also called *profibrinolysin)* is activated to plasmin by activators present in the tissues, serum, urine, and some bacteria. Normally fibrinolysis is prevented by the presence of *anti-plasmin* in the blood. This remains attached to the plasma albumin.

Importance of Coagulation of Blood: The phenomenon of coagulation is of great physiological significance. Its purpose is to stop further haemorrhage.

When bleeding occurs, the shed blood coagulates and bleeding vessels become plugged off by the clot. The retraction of the clot compresses the ruptured vessels further and in this way bleeding is stopped.

Relation between Blood and Lymph

Blood is a fluid connective tissue circulating in the body. It provides one of the methods of communication between the cells of different parts of the body.

The blood carries oxygen and nutrients to the tissues and carbon dioxide and waste products from the tissues to the excretory organs. In addition it conveys antibodies to the site of injury or disease and hormones or chemical messengers from the endocrine glands to various target organs. The blood is the medium of transportation of all these substances.

The blood constitutes approximately 8 per cent of the body weight or 5.6 litres in a 70 kg man.

Composition and Functions of Blood: The blood is composed of a faintly yellow transparent fluid known as the *plasma* and floating in this fluid are numerous *cells* or *corpuscles* of different kinds. The plasma constitutes approximately 55 per cent and the corpuscles about 45 per cent of the volume of blood.

I. The Cellular Contents of Blood

There are three varieties of cells or corpuscles present in blood.

A. *Erythrocytes or red blood corpuscles.*

B. *Leucocytes or white blood cells.*

C. *Blood platelets.*

Erythrocytes: As a matter of fact the word 'erythrocyte' is inappropriate for mammals because the adult corpuscle lacks a nucleus and, therefore, is not a complete cell. However, the absence of nucleus in mammals increases respiratory efficiency of the red blood corpuscle. All vertebrate, other than mammals have nucleated red cells.

Shape: These are circular, biconcave non-nucleated discs. The central part of the corpuscle is much thinner than the circumference, thus the term biconcave. In edge view the outline is like that of a dumb bell, this shape favours flexibility and the absorbing and releasing of gases quickly. RBC's of camel are oval in shape.

Size: In man, the living corpuscle is 8.5 μ in diameter. Rabbit has slightly smaller corpuscles than man. The extremes are 9.4 μ in elephant and 2.5 μ in musk deer. Amphibian RBC's are largest among the vertebrates.

Number: The normal erythrocyte count is usually slightly lower in women than in men.

Women : 4.5 to 5 million per cubic millimetre of blood.

Men : 5 to 5.5 million per cubic millimetre of blood.

The number varies with different pathological and physiological conditions. For example, the number is decreased in pernicious anaemia. On the other hand, continued living at high altitudes or exposure to carbon monoxide causes an increase in the number of circulating red blood corpuscles. An abnormal rise in RBC count is called *polycythemia*.

Colour: A single fresh corpuscle is pale-yellow, but when the corpuscles are seen in a mass, the haemoglobin gives the characteristic red colour.

Structure: The erythrocytes are surrounded by a thin peripheral membrane (0.02 μ) consisting of a lipid-protein complex.

It serves as an elastic and semi-permeable barrier. The nucleus and cytoplasmic organelles like endoplasmic reticulum, golgi apparatus, centrioles and mitochondria were lost in the later stages of development. The internal cytoplasm is a homogeneous colloidal sol.

Physical Properties: The red corpuscles are soft and flexible. They withstand much bending, squeezing and deformation as they are pushed through the narrow capillaries.

Rouleaux are columns of corpuscles, stacked one above the

other like a pile of coins. The adherence is due to surface tension. These are formed spontaneously in a stagnant circulation or in a column of drawn blood.

Crenation is a shrivelling of the corpuscles due to concentration of the plasma by evaporation. It produces a prickly contour.

Haemolysis is the outward passage of haemoglobin from the corpuscle due to injury of plasma membrane caused by lipid solvents, bile salts and snake venom. *Agglutination* is the clumping of corpuscles induced by various agents like salts solutions, glucose solution and antibodies.

Chemical Composition: The respiratory pigment of red corpuscles is *haemoglobin*. It is a complex protein having iron as one of the main constituents.

The membrane of the erythrocyte is composed of 49 per cent protein, 44 per cent lipid, and 7 per cent carbohydrate. 26.5 crore molecules of haemoglobin are packed in the intracellular frame-work of a red corpuscle.

Formation (Erythropoiesis): The erythrocytes are formed in liver, spleen and lymphs in the *embryo* ; and in the red bone marrow (myeloid haemopoietic tissue) in the *adult*.

The red bone marrow is and in the cancellous bone at the extremities of long bones and between layers of compact bone in flat and irregular bones such as the cranium, vertebrae, ribs, sternum, clavicles, scapulae and Pelvis. The process is known as *erythropoiesis* and is completed in 72 hours.

The development of a mature red corpuscles takes about a week, during which the endothelial cell enlarges, divides, forms haemoglobin and finally loses its nucleus. This process is called maturation and takes place along the following lines:

Stems cells or Myeloblasts ⟶ Proerythroblasts ⟶ Erythroblasts ⟶ Normoblasts ⟶ Reticulocytes ⟶ Erythrocytes.

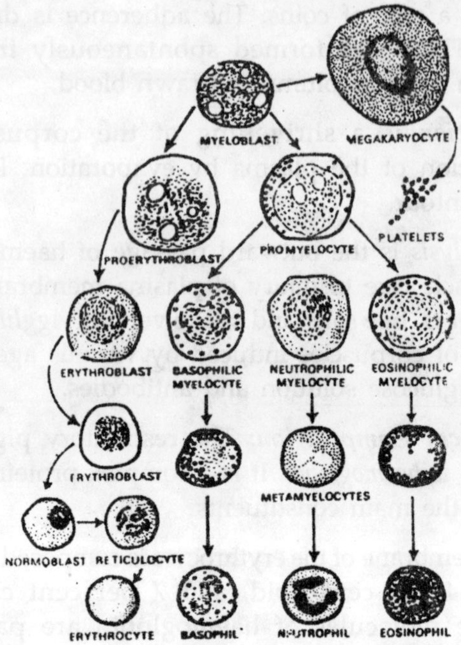

Erythropoiesis.

Myeloblast is an amoeboid cell with fairly abundant cytoplasm and a relatively primitive, undifferentiated nucleus.

Proerythroblast is larger than the myeloblast. Its nuclear network is slightly coarser and the cytoplasm is deeply basophilic due to the presence of ribonucleic acid (RNA).

Erythroblasts are half the size of the proerythroblasts due to mitosis in that stage. The nucleus is checkered with coarse chromation masses. The cytoplasm is losing its RNA and is acquiring haemoglobin.

Normoblasts are derived from erythroblasts by mitosis, hence they are smaller in size. The nucleus becomes progressively smaller and mitosis ceases after this stage. The cytoplasm stains strongly acidophilic due to haemoglobin.

Reticulocytes are young, immature erythrocytes with a delicate network of ribonucleic acid in the cytoplasm. The nuclei of normoblasts have been lost by extrusion in this stage.

Erythrocytes are mature, enucleated cells which are also called *red blood corpuscles.*

The normal maturation of a red blood corpuscle requires the presence of a number of different chemical substances:

(a) Vitamin B_{12} (Cynacobalmin or erythrocyte-maturing factor) and folic acid are necessary for the development of the proerythroblast successively into erythroblast and normoblast.

Vitamin B_{12} is present in food (mainly animal proteins) but its absorption from the small intestine is dependent upon a mucopolysaccharide, the *intrinsic factor,* secreted by the parietal cells of the gastric mucosa.

(b) Iron is necessary for the provision of haemoglobin to fill the immature erythrocytes before they can become the mature red corpuscles. Lack of iron in the diet or loss of iron in bleeding causes iron-deficiency anaemia. The normal haemoglobin levels in the blood are as follows:

Men	14 g per 100 ml blood
Women	12 g per 100 ml blood

(c) In addition to a normal diet, containing protein and ran small amounts of cobalt, copper, nicotinic acid, riboflavin and Vitamin C are also essential for erythropoiesis.

The development of red blood corpuscles is controlled by a feed-back mechanism. Deficiency of oxygen following haemorrhage or because an individual, lives at high altitudes where the oxygen pressure in the atmospheric air is reduced; and a hormone called *erythropoietin,* secreted by the kidneys are the two main factors that stimulate the bone marrow to increase its production of erythrocytes.

Metabolism: The mature mammalian erythrocyte lacks mitochondria ; hence the cytochrome system is absent and the tricarboxylic acid cycle is not evident. The energy of the mature erythrocyte is supplied primarily by anaerobic glycolysis and the phosphogluconate pathway.

Life Span and Disposal: The Life span of human erythrocytes is approximately 120 days (100 days in frog). When worn out, the cells break down in the spleen, which, however, saves and stores the iron for use in the preparation of fresh haemoglobin. The protein part of the erythrocyte is converted to a substance called *biliverdin* which is almost completely changed to *bilirubin,* a yellow pigment. While circulating in the blood, bilirubin is bound to the plasma protein albumin. It is carried by splenic and portal veins to the liver where it is again changed into a conjugated form arid excreted in the bile as bile pigment.

Functions: Haemoglobin in the erythrocytes carries oxygen from the lungs to all the cells of body for tissue respiration.

Leucocytes or White Blood Cells: The white blood cell or leucocyte contains a nucleus and is therefore, a true cell.

Structural Characteristics: Leucocytes are typical cells with amoeboid ability. There is a centriole near the nucleus but mitosis does not occur in the adult state. As in all true cells; mitochondria and Golgi apparatus is present.

The leucocytes differ from the erythrocytes in that they possess a nucleus, are fewer in number and larger in size. They vary in size from about 8 to 15 microns.

Number: The average normal limits are between 6,000 to 10,000 per cubic millimetre of blood. The count in children is higher than in adults. In pathogenic states, there are variations from the normal number.

An increase in the number of white blood cells is called leucocytosis: more than 12.000 per cu. mm. indicates some disease. This increase is out of proportion for one or more types.

(i) *Neutrophils* increase in pus-forming infections.

(ii) *Eosinophils* increase in allergies and parasitic infections.

(iii) *Basophils* increase in Chicken-pox.

(iv) *Lymphocytes* increase in whooping cough.

(v) *Monocytes* increase in tuberculosis.

A decrease below 6,000 is called leucopenia as in typhoid fever.

Classification: The leucocytes are divided into two main varieties:

(i) Granular or Polymorphonuclear leucocytes,

(ii) Agranular or Mononuclear leucocytes.

This classification is based on:

(i) The shape of the nucleus

(ii) The presence or absence of granules in the cytoplasm and

(iii) The type of granules, if present.

Granular Leucocytes and their Application

These cells develop in the red bone marrow from the same parent cells as the erythrocytes., *i.e., myeloblasts* in the red bone marrow. Before entering the blood stream, they develop through promyelocyte and myelocyte stages.

Myeloblast → Promyelocyte → [Basophilic myelocyte / Neutrophilic myelocyte / Eosinophilic myelocyte] Metamyelocytes

→ Juvenile → Basophil
→ Juvenile → Neutrophil
→ Juvenile → Eosinophil

As these cells develop, specific granules appear in larger numbers and they retain their nucleus. The granulocytes constitute about 75 per cent of total white Wood cells and three varieties are described:

(a) Neutrophils approximately 70%

(b) Eosinophils approximately 4%

(c) Basophils approximately 1%

Neutrophils (Phagocytic): The cytoplasm is closely packed with fine, inconspicuous granules that absorb both acid and alkaline dyes thus producing a neutral purple colour. The nucleus has several (3 to 5) lobes connected by thin chromatin strands. Fewer lobes indicate less mature cells. About 3 per cent of neutrophils from a female show the sex chromatin attached to one of the nuclear lobes by a thin stalk forming the so-called "drumstick chromosome".

Functions: Neutrophils protect the body against the invasion of bacteria. They are attached in large numbers to any area of the body which has been invaded by micro-organism. Neutrophils leave the blood by squeezing themselves out through the walls of the capillaries in the infected area. This process is known as *diapedests.* Thereafter they ingest and kill the organisms by digesting them by means of various enzymes, a process known as *phagoctyosis.*

The *pus* which may exude from an infected area consists of destroyed tissue, live micro-organisms and dead neutrophils which have ingested more micro-organisms and dead neutrophils which have ingested more micro-organisms than they could digest.

The life span of neutrophils which remain in blood vessels is about 30 hours. After migrating from the blood capillaries into the inflamed tissue, they play a defensive role there and soon die.

Eosinophila (Non-phagocytic): The diameter of the cell is nearly twice that of a red blood corpuscle. The cytoplasm is packed with coarse round granules which absorb an acid dye such as *eosin* which is red. The nucleus is usually bilobed, with a connecting isthumus.

Functions: Eosinophils engulf the particles which are formed when antigens and antibodies react with each other. In allergic conditions, such as asthma and parasitic infections of the digestive tract, they are increased in numbers.

Basophils (Non phagocytic): The cytoplasmic granules absorb on alkaline or basic dye such as *methylene blue* and are stained dark blue to purple in colour. The granules are not as numerous as in eosinophils, but they are large, spherical and almost cover the nucleus which is generally S-shaped.

Function: Basophils are actively amoeboid and ingest small particles like carbon.

Agranular Leucocytes and their Application

They have a few non-specific or on granules in the cytoplasm and the nucleus is spherical to kidney shaped. They comprise about 25 per cent of all leucocytes, and have two varieties: (a) Lymphocytes, large and small, (b) Monocytes.

(a) Lymphocytes (Non-phagocytic)

The large and small lymphocytes closely resemble each other and differ only in size. *Small lymphocytes* are most common in the blood. They are relatively larger than red blood corpuscles. A small lymphocyte has a relatively large nucleus, surrounded by a narrow rim of basiphilic cytoplasm. The nucleus, is nearly spherical, having a slight indentation. It is very dense and dark because of heavy chromatin masses.

Large lymphocytes are primitive, proliferative cells residing in the lymph nodes that give rise to the active small, lymphocytes. The nucleus in generally kidney-shaped.

Function: The lymphocytes are produced in the lymphatic issue of the body which is present in the spleen, lymph nodes thymus, tonsils and more scattered nodular masses.

(i) Lymphocytes are concerned with the process of immunity. They produce serum globulin β and γ-globulins), one of the plasma proteins, which gives rise to *antibodies* and *antitoxins.*

(ii) They contribute to scar-formation after injury and thus facilitate wound-healing.

(iii) Lymphocytes play an important role in the immunological reactions to tissue transplantation.

Monocytes (Phagocytic): They are largest of the leucocytes and are relatively few in number. The nucleus varies from oval, round or indented to kidney shape and stains lightly than that of the lymphocytes. It tends to be eccentric in position. The abundant cytoplasm is lightly basiphilic.

Function: The monocytes originate from a system of primitive cells, the *recticido endothelial system* which is found in organs such as the liver, spleen, lungs and lymphatic glands.

Their function closely resembles that of the neutrophils in that they are actively motile, phagocytic in action and will leave the blood capillaries to ingest micro-organisms and other foreign material that may be introduced into the tissue.

They are the prime scavangers of cells and tissue debris. They play a vital role in removing damaged tissue and thus preparing the way for regenerative processes of the body.

Blood Platelets: These are protoplasmic disks that are found in mammalian blood (lower vertebrates have spindle-shaped cells named thrombocytes). Platelets arise as detached tips of protoplasmic processes extending from the cytoplasm of giant cells *megakaryocytes,* of red bone marrow. The shape is oval to round, often stilette. They are 2-3 μ in diameter. The protoplasm is granular and deeply basophilic in the centre but is pale and homogeneous on the periphery. There are approximately 300,000 platelets in a cubic millimetre of blood.

Functions: Agglutinated platelets are associated with blood, clotting, both inside and outside of blood vessels.

Liquid Portion of Blood (Plasma)

The plasma is composed of the following constituents:

1. *Water* 90 to 92 per cent.
2. *Plasma Proteins:* Proteins in the plasma perform several important functions:

(a) They exert an osmotic force of about 25 mm of mercury across the capillary wall which is called *oncotic pressure.*

(b) They give viscosity to the blood, thus assisting to some extent in the maintenance of the *blood pressure.*

There are four principal plasma proteins:

(i) *Serum albumin:* It is formed in the liver.

(ii) *Serum globulin:* This is derived from the lymphocytes.

(iii) *Fibrinogen:* This protein is produced in the liver and is necessary for blood clotting. Plasma defibrinated by clotting or whipping is called *serum.*

(iv) *Prothrombin:* It serves as factor II in the blood clotting mechanism and is formed in the liver.

Mineral Salts (Inorganic Salts): These include sodium chloride and sodium bicarbonate. ' Small amounts of potassium, magnesium, phosphorus, calcium, iron, copper and iodine are also found in the plasma. Mineral salts are necessary for the *formation of cells, the contraction of muscles, the transmission of nerve impulse* and to maintain *the balance between acids and alkalis* in the body. In health, the blood is always *slightly alkaline* in reaction (pH 7.4).

Nutrient Materials: After the digestion of carbohydrates, proteins and fats, *amino acids, glucose, fatty acids, glycerol* and *vitamins,* etc., are absorbed into the blood plasma from the alimentary tract. These are the nutrient materials which are required to maintain the body functions.

Organic Waste Products: *Urea, uric acid* and *creatinine* are waste products of protein metabolism which are formed in the liver. They are carried by the blood plasma to the kidneys for excretion.

Hormones: The "chemical messengers" secreted by the endocrine " glands are passed directly into the blood which transports them to the target organs.

Enzymes: Some enzymes are also present in the plasma.

Antibodies: These are protective substances consisting of complex proteins that are produced in the spleen and lymph glands by a special type of cells *plasma cells*. The antibodies are produced in response to the entrance of a foreign agent, the *antigen* into the body.

Gases: Small amounts of oxygen, carbon dioxide and nitrogen are dissolved in the plasma.

Summary of Blood Functions: The blood performs many important and vital functions in the animal body and is often described as "the river of life". Its various constituents carry out different functions as given under:

Erythrocytes

(1) *Transport of* O_2: The red corpuscles carry oxygen (in combination with haemoglobin) as oxyhaemoglobin from the lungs to the tissues for oxidation of food to release energy for various metabolic activities.

A small amount of CO_2 is also transported by erythrocytes from the tissues to the lungs as carbamino-haemoglobin for its disposal.

Leucocytes

(2) *Fighting Diseases:* The leucocytes act as soldiers and scavangers of the body.

(i) The *neutrophils* are actively amoeboid and phagocytic. They appear in large numbers in and around the infected tissues and engulf live and dead bacteria, fungi and viruses.

(ii) The *monocytes* are the prime scavangers of injured and dead cells. They devour these elements to clean the body.

(iii) The *basophils* engulf particles of dust, smoke, etc.

(iv) The *lymphocytes* partly manufacture beta and gamma globulins (antibodies) of the plasma. They also contribute to the scar formation after injury and thus facilitate in wound-healing.

(v) The *eosinophils* flock to the digestive and respiratory treats in allergies.

The dead leucocytes damaged by bacteria together with fragments of tissues form the pus.

Blood Platelets

(3) The blood-platelets are associated with blood-clotting.

Plasma

(4) *Transport of Nutrition:* It carries glucose and other organic metabolites from the alimentary canal and liver to various parts of the body for energy, growth and repair.

(5) *Disposal of $CO_{2:}$* Carbon dioxide is produced in the body as a metabolic waste. The blood plasma picks up CO_2 from all the tissues and brings it to the lungs for expiration.

(6) *Transport of Excretory Products:* Plasma carries nitrogenous waste products from various tissues to the liver where they are converted to urea and other harmless products.

The latter are transported from the liver to the kidney for removal as urine.

(7) *Distribution of Hormones and Vitamins:* It distributes hormones from the endocrine glands and vitamins to the specific place of their action in the body.

(8) *Regulation of Water-balance:* As the plasma supplies water to the tissues and receives excess water formed in metabolic processes. It regulates the water balance of the animals. Water provides vehicle for other constituents whereas its excess it eliminated through kidney.

(9) *Osmotic Balance:* Plasma proteins and various mineral

ions in the plasma maintain osmotic balance, pH balance, buffer-capacity and have varied effects on tissue cells.

(10) *Protection against Diseases:* Immunoglobulins of the plasma act as antibodies that neutralise the harmful effects of foreign agents in the body. They also form basis of blood types.

(11) *Regulation of Body Temperature:* Plasma regulates the body temperature by carrying heat from heat producing organs like muscles to those parts where heat is lost, *e.g.*, lungs (via expiration).

(12) *Blood-clotting:* Fibrinogen, prothrombin and other blood-clotting factors present in the plasma, participate in blood clotting, thus helping against haemorrhage.

(13) *Lymph Formation:* It participates in the formation of lymph and keeps all the tissues moist.

Significance of Blood Groups

The membranes of human red cells contain a variety of antigens called agglutinogens. Chemically, they are glycolipids and are inherited as Mendelian dominants. The most important and best known of these are the A and B of agglutinogens which constitute the ABO system of blood groups. According to this system, the individuals are divided into 4 major blood types, types A, B, AB, and O, on the basis of the agglutinogens present in their red cells. Besides, the other well-known blood group systems are Rh system, the McKell, lutheran, Duffy, Diego and Kidd systems.

Blood Groups Antibodies: Antibodies against agglutinogens are called agglutinins. Antibodies are produced mainly in the bone marrow and in the spleen, liver and lymph nodes. The blood group antibodies belong to the serum globulins and may be divided into two broad classes:

(*i*) *Immune* and

(*ii*) *Naturally occurring.*

Immune Antibodies: They may be produced by exposure to the red cells of another individual. This exposure may occur via (i) a transfusion or (ii) during pregnancy, when foetal red cells cross the placenta and enter the circulation of the mother.

(i) If the blood of a person containing Rh antigen is transfused into a person who lacks this antigen, the recipient is said to be immunised and the corresponding antibody (Rh-antibody) will appear in his serum.

(ii) Red cells from a foetus carrying an antigen. Inherited from its father but absent in the mother may also produce the equivalent immune antibodies by entering the maternal circulation through the placenta.

Immune blood group antibodies belong to the category of γ G globulins (IgG). Their molecular weight ranges from 150,000 to 160,000 and they have a unique ability to cross the placenta.

Naturally Occurring Antibodies: These antibodies are present in the serum of an individual who has never been immunised by the equivalent antigen. Such antibodies are first to be formed in a new born animal or human. Example are anti-A and anti-B agglutinins..

Naturally occurring antibodies are usually γ M globulins (IgM). They have a molecular weight of about 900,000 and do not cross the placenta.

The antigen-antibody relationship is one of great specificity like that between lock and key.

3

Behaviour in Feeding

In the whole animal kingdom, feeding behaviour is incredibly varied and entirely very complex. In those animals which feed almost continuously, there is a complete spectrum of temporal for example certain filter feeding invertebrates to others, like some parasitic insects and carnivorous vertebrates that feed rarely, at intervals of many hours, days or weeks. In some animals, when there is no feeding at all, their life cycle exhibits phases, such as in the adult of some insects and in the male fur seal while he is ashore defending his territory. Most of the animals which feed at relatively short intervals measured in minutes or a few hours exists somewhere in between these extremes. Rats, mice, birds, fishes are flies and included in most of the examples and we can restrict our discussion to those animals which are restricted to any of the above animals which is more familiar to the conditions.

In rats, feeding occurs at an average interval of about 2 to 4 hours, and this interval could be changed resulting into wide variations and individual difference by training. Nevertheless, a cyclic pattern of one kind or another persists, with statistically predictable intervals between feeding bouts. Among Cattle feeding occurs with a regular pattern of distinct bouts in one complete day. In Dolichonyx oryzivorus, the bololink, a small passerine bird feeding occurs with a rhythmical pattern.

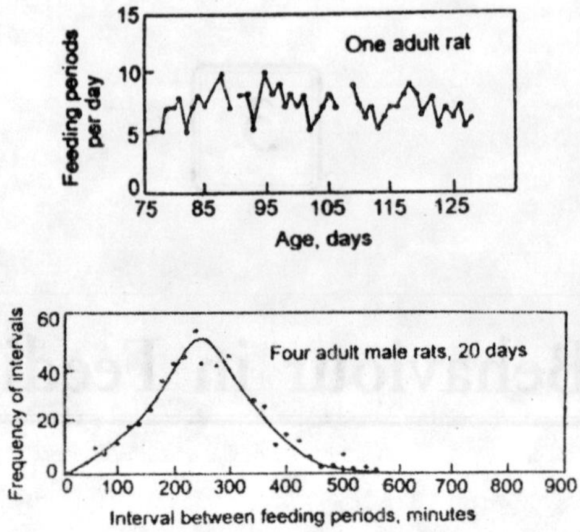

Rhythmical expression of the feeling behaviour recurring feeding in different ways.

By expanding the time scale and observing the behaviour exhibiting one cycle of feeding activity, more details of the pattern could be revealed. Generally, locomotion is the prelude to a bout of eating either in free exploration of the environment or moving in such a location where food is found habitually. This appetitive phase is followed by the behaviour associated with selecting, preparing and ingesting food, *i.e.*, the consumatory phase. There are a variety of other activities like resting sleeping, drinking, social encounters and locomotion which occupy the intervals occurring in the above phase. Cattle spend most of its time in ruminating, *i.e.*, regurgitating food, chewing it and swallowing it again in between bouts of feeding.

Feeding Mechanism

On the basis of the primary feeding strategies, animals can be grouped into the following classes:

(i) *Autotrophs:* It includes those organisms which synthesise the essential organic compounds from inorganic substances. Thus these organisms feed independently. Autotrophs, not only include plants but euglena, volvox and few bacteria are also said to feed independently.

Depending upon the way of producing energy, organism could be classified into two groups:

(i) Photosynthetic autotrophs (ii) Chemosynthetic autotrophs.

Photosynthetic Autotrophs: Those organisms which convert solar energy of sun into chemical energy by a process of photosynthesis are known as photosynthetic autotrophs. For example, plants, euglena, volvox. They are able to synthesise their own amino acids, fats, sugars, etc. from inorganic substances like CO_2, nitrates, etc.

Chemosynthetic Autotrophs: Those organisms which manufacture their all protoplasmic requirements from inorganic substance through a process called chemosynthesis are called chemosynthetic autotrophs. For example sulphur and iron bacteria (also called chemotrophs) which use the energy from oxidation of simple inorganic chemicals like iron and sulphur.

(II) *Heterotrophs:* It includes animals and some non-green plants which can neither utilise solar energy nor can synthesise organic substances from inorganic material. (CO_2, nitrate, etc.) Heterotrophs need organic substances for their life.

There are several ways by which heterotrophs obtain their food from plants and animals:

(i) *Holozoic Nutrition:* In this type of nutrition organism obtain energy by oxidising pre-existing organic compound of plants or animals. There are a number of ways by which they feed, like:

Herbivores: feeding upon plants and vegetation for example, cow, horse, rabbit, buffalo, etc.

Carnivores: feeding upon other animals or on flesh, for example, lion, tiger, dogs, etc.

Omnivores: Feeding upon any type of food directly upon both plants and animals, for example, human beings, flies, cockroaches, etc.

Cannibulus: Feeding upon its own species. Although it is not a common method of taking food but still it is usually seen to be exhibited by fishes. It is under the direct control of nature.

Detritus: Feeding upon organic matters present in soil or mud. For example, earthworm.

Voraceous: Feeding upon crustaceans and larvae. For example, gambusia fish, gastropods, etc.

Predators: Feeding upon lowei species especially on invertebrates, for example, birds like vultures, cranes, etc.

Insectivores: Feeding upon insects, for example, frog, bat, lizards, etc.

Ant Eaters: Feeding specially upon ants due to adaptations, for example, echidna (spiny ant eater), manis (scaly ant eater), etc.

Food Robbers: Feeding upon fully or semi-digested food material of the host from its alimentary canal, for example, ascaris, taenia, etc.

Saprozoic: Deriving food from dead decaying matter. They are also called scavengers. For example, euglena, paramecium.

(ii) *Saprophytic:* Those organisms which take dissolved decaying organic material from their environment are called saprophytic. For example, fungi, amoeba, euglena, etc.

(iii) *Osmotrophic:* Those organism which feed upon pre-digested food material from the host by diffusion through their body wall are called osmotrophic. For example, trypanosoma, taenia solium, etc.

(iv) *Sanguivores:* Those organisms which feed upon blood of other animals are called Sanguivores. In these organism an anti-coagulin like heparin, hirudin, lamphredin, etc. is secreted in order to prevent clotting of blood in their sucking apparatus. For example, leeches, lampreys, mosquitoes, etc.

(v) *Parasitic:* Those organisms which feed at the expense of living organisms which entertain them as "host" normally without any fatal results. For example, ascaris, liver fluke, head louse, tapeworm, etc.

(vi) *Symbionts:* In this two organisms exhibit a sort of mutual association where both of them are benefited.

Food Preferences: No two species living together at the same time and place eat exactly the same staple food. All animals are selective in choosing food in their natural habit, some more so than others. According to Lack (1954) and Mayr (1963) this is the competitive principle. Thus selectivity of food is a matter of major biological importance, not only because it has a bearing effect on nutrition but also because of its relationship to problems of interspecific competition. According to Hafez and Schein (1962), when animals are given a choice they show preferences. "Cattle exhibit preferences not only for certain plant species but for the same species at different stages of growth and even for the various parts of an individual plant and for individual plants within a species". Among the seed eating birds (Morris 1955, Wood-Gush 1955, Kear 1962), rats (Barnett 1963) and primates (Harlow and Meyer 1952, Berkson 1962), preferences are equally clear.

Feeding in Group-living Herbivores: The animals which live in huge groups which together make up some 90 per cent of the total weight of mammals living on the serengeti includes buffalo, zebra, wildbeest, topi, and Thomson's gazelle. These five in the just sight appear to live on the same species of grass, herbs and small bushes, however, the appearance is illusory. When the stomach contents of all of these except buffalo was studied by Bell and his colleagues, it was found that each species was living on a different part of the vegetation. These different parts of the vegetation differ in their food qualities lower down these are succulent, nutritious leaves, higher up are the harder stems Highly nutritious fruits are also found distributed sparsely. Bell and his colleagues found that only the Thomson's gazelle eats much of these nutritious fruits.

The other three species differ in proportions of lower leaves and higher stems that they eat. The zebra eats mostly stem material, the wildbeest depends on the leaves and the topi is intermediate.

In order to understand the different feeding preferences, we can consider two things firstly to study their digestive system and secondly their body sizes. The digestive system can be differentiated into the non-ruminants (the zebras) and the ruminants (wildbeest, topi, and gazelle which are like the cow). Non-ruminants are not able to extract much energy from the hard part of the plant; however, this is more than made up for by the fact that food passes much more quickly through their guts.

Thus, when there is only a short supply of poor quality food, the wildbeest, topi and gazelle enjoy an advantage. They are ruminants and have special structures in their stomaches (the rumen), containing special micro-organisms which can break down the hard parts of the plants. Food passes only slowly through the ruminant's guts because ruminating, digesting the hard parts, takes time. The ruminant continually regurgitates food from its stomach to its mouth to chew it up further (that is, what a cow is doing when 'chewing cud'). Only when it has been chewed up almost to a liquid can the food pass through the rumen, and on through the gut. Larger particles cannot pass through until they have been chewed down to size.

Therefore, when food is in short supply, a ruminant can last longer than a non-ruminant because it can extract more energy out of the same food. The difference can partially explain the eating habits of the serengeti herbivores. The zebra chooses area where there is more low quality food. It migrates first to unexploited areas and chomps the abundant low quality stems before moving on. It is a fast in/fast out feeder, relying on a high throughout of incompletely digested food. By the time the wildbeest (and other ruminants) arrive, the grazing and trampling of the zebras will have worn the vegetation down. As the ruminants then set to work they eat down to the lower, leafier parts of the vegetation. All of which fits in with the differences of stomach contents with which we began.

Body size includes the other parts of the explanation. Larger animals require more food than smaller animals, but smaller animals have a higher metabolic rate. Smaller animals can, therefore, live where there is less food, provided that it is of high energy content. That is why the smallest of the herbivores, Thomson's gazelle, lives on fruit, which is very nutritious but too thin on the ground to support a larger animal. By contrast, the large zebra lives on the masses of low quality stem material.

The differences in feeding preferences lead, in turn, to differences in migratory habits. We have seen that wild beast follow, in their migration, capricuous pattern of local rainfall. The other species do likewisé, but when an area is fuelled, by rain, for exploitation, the mammals migrate towards it in an orderly pattern. The larger, less fastidious feeders, the zebras, move in first, the choosier, smaller wildebeest come later and the smallest species of all, Thomson's gazelle, arrive last. The later species depend on the preparations of the earlier, for the action of the zebra fits the vegetation for the stomaches of the wildbeest and gazelle. So if we are to understand the feeding habits of the species, we must consider it in relation to the whole ecology of the species, and its relation to other species. Behaviour is an inseparable part of a whole system, made up of body size, gut morphology and the habits of associated species.

Carnivores and their Characteristics

Among the mammalian carnivores, one member of the family Felidae and several members of the family Canidae have evolved complex social behaviour that is related to cooperative capture of prey. All of these species have been the subjects of intensive behavioural and ecological study in their natural habitats.

Lions: The lion (Panithero leo) lives in closed social unit and is most abundant in the grasslands and open woodlands of Africa. Schaller has studied the behaviour and ecology of this species, its competitors and its prey.

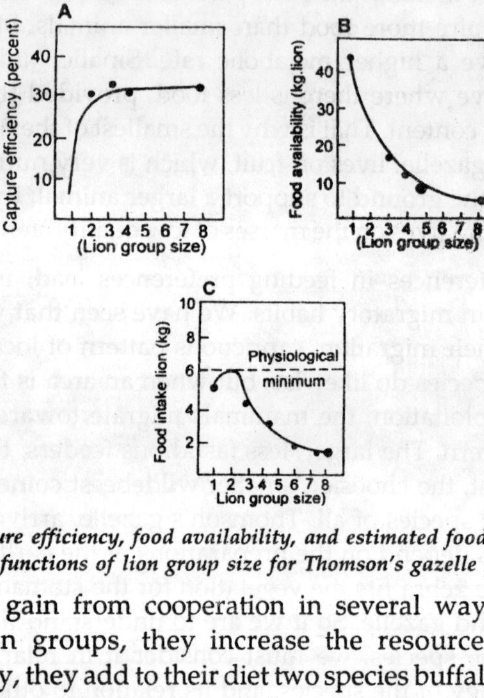

Capture efficiency, food availability, and estimated food intake as functions of lion group size for Thomson's gazelle prey.

Lions gain from cooperation in several ways. First, by hunting in groups, they increase the resource spectrum; specifically, they add to their diet two species buffalo (Syncerus caffer) and giraffe (Giraffa camelopardalis) that an individual lion could never attack alone. Second, cooperative hunts are at least twice as successful as solitary ones. Third a group can consume a captured prey more fully than can an individual. Fourth, the group can drive other predators and scavengers from the food Schaller found that plain dwelling prides kill less than half their food, relying on other predators, such as hyenas (Crocuta crocuta), to make kills for them. Although lions hunt in groups, the amount of cooperation is not extensive.

When the lions encounter a herd of prey, such as zebra (Equus burchelli), the pride female fan out, sometimes encircling the herd. Lions are not endurance runners and rely on a stalk and rush tactic. Although lions are more effective when hunting upwind. Schaller found no evidence that they do so more often than would be expected by chance.

Once a kill has been made, the males, which are larger than the females but which do little hunting, may drive the females from the kill. The cubs are the last to get food. Schaller believes that lion populations are regulated directly by food supply through starvation of cubs. The reproductive rate remains constant, but the mortality rate of the young can be very high, exceeding 80 per cent when food is scarce. Not in all cub, mortality is the result of starvation; strange males that move into a group may kill cubs.

Caraco and Wolf (1975) used Schaller data to relate ecological factors to lion foraging group size. When hunting small prey, such as Thomson's gazelles (Gazella thomsoni), two lions are more than twice as efficient as one, but three or more do no better than two. Since the available food for each lion per kill decreases with increasing group size, we would expect foraging groups of about two. Caraco and Wolf went on to show only pairs of lions can take in the minimum daily amount of food if they feed exclusively on Thomson's gazelles; Schaller reported that the mean foraging-group size ranged from 1-5 to 2, close to expected. In hunting larger prey such as zebras and wildebeest (Connochaetes taurinus), a foraging group size of 2 was again the most efficient, but groups of 1 to 4 could still attain their minimum daily requirement. However, Schaller found larger than expected foraging groups of about 4 to 7 Caraco and Wolf were forced to suggest that factors other than prey size, such as the higher reproductive success documented for larger prides (lionesses share in the feeding of cubs), could influence group size.

African Wild Dogs: These are unlikely big game predator weighing only about 18 kg (40 lbs), typically catches prey weighing as much as 250 kg (Schaller 1972). The dogs live in mixed-sex packs that average ten adults and, when there are young in a den, range out from it to hunt. The pack is tightly organised. Of particular interest here is the prehunt ceremony, in which dogs draw back their lips to expose their teeth, nibble and lick each other's mouths, and run winning from pack member to pack member.

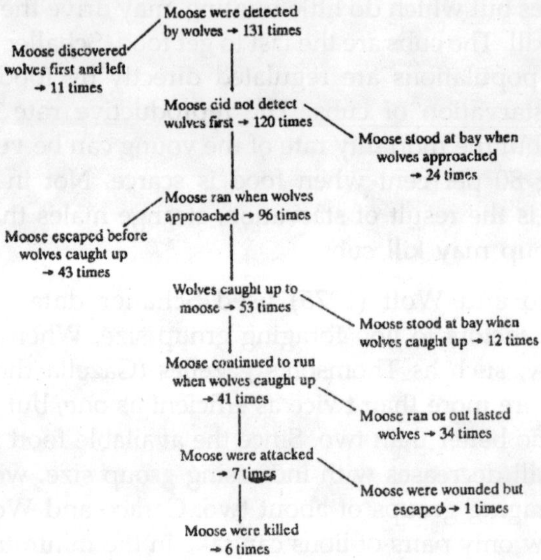

The greeting seems to represent ritualised food begging. Setting out on a hunt, the dogs travel at a trot and fan loosely over the terrain; certain adults consistently lead some packs, but in other packs there seem to be no leader. Schaller (1972) describes the hunt in his field notes:

Moose-Wolf Relation

For African wild dogs, the hunting success is extremely high; say about 90 per cent of all hunts result in the capture of at least one prey item.

Wolves (Canis lupus): It is a social carnivore of northern temperate and arctic regions. It typically preys on deer

(Odocoileus spp.), moose (Alces americana), buffalo (Bison bison), sheep (Ovis spp.), Caribou (Rangifer tarandus) and elk (Cerous canadensis). Pack size varies greatly (Mech 1970), but most have fewer than eight members. Wolves hunt by·travelling over their territory after consuming their previous kill. Most often they use scent; the sea animals stop when they detect the odour of prey. The group may stand nose to nose with tails wagging for a few seconds, and then all follow the leaders directly towards the prey. They may also use chance encounter and tracking. If they detect the prey at some distance, they proceed with restraint and stalk sometimes to within 10 metres of the prey. When the prey detects the wolves, it may stand its ground or flee. Once the prey is in flight, the wolves rush. Wolves must get close to their prey during the stalk and rush on they will quickly give up; if they cannot make an attack, they may chase the prey, usually for less than half a mile. Some earlier studies described wolves as coursers, but Mech has concluded that they are mainly rushers and stalkers.

The musk ox (Ovibos moschatus), one of the prey of wolf, defends itself in an unusual way. Adults form a circle, facing outward and protecting the young inside. There is little evidence that wolves can break this defence, in areas where musk oxen are present, wolves eat mainly other prey or capture an occasional post-reproductive individual.

When hunting moose, which are solitary, wolves rarely attack a prime individual (between one and six years old). Most studies of wolf predation have indicated that they are highly selective of young, old, and sick prey. Although earlier reports described the main killing tactic as hamstringing, wolves actually avoid the hooves and direct bites at the rump, flanks, shoulders, neck and nose.

Significance of Anti- Predator Behaviour

As we know that most of the species are preyed upon by animals of at least some other species. Thus in survival and reproduction, avoidance of predation plays an important role.

There are certain defensive mechanism which are operated throughout much of the life of an individual but others are used only when a predator is detected or when a predator attacks the individual. Edmunds (1947) refers to two types of mechanism, primary and secondary. Following Kruuk (1972) Edmunds define primary defence mechanism as those which operate regardless of whether or not there is a predator in the vicinity. The behavioural components of such mechanism may occur before any predator is detected. The mechanism include :

(1) Hiding in holes;
(2) The use of crypsis;
(3) Mimicking inedible objects; (4) exhibiting a warning of danger to predators;
(5) Mimicking individuals in category;
(6) Timing activities so as to minimise the chance of detection by a predator;
(7) Remaining in a situation where any predator attack is likely to be unsuccessful because of possibilities for secondary defence;
(8) Maintaining vigilance so as to maximise the chance of detecting the advent of a predator.

All of these mechanism decreases the chance that a predator attack will occur, many of them have a considerable influence upon habitat selection. For some species which are very vulnerable to predation, the avoidance of predators limits the distribution of the animals even more than do factors related to body maintenance and feeding. The selection of a suitable habitat may be accomplished by the recognition of a factor which is not in itself advantageous but which is an indicator of the presence of advantageous features, such as those which make possible feeding or predator avoidance. For example, a bird, such as a nightjar, which is camouflaged when amongst dead wood might fly over open downland towards a dark green mass of woodland because suitable hiding places are likely to be found around the edge of the wood.

Edmunds defined secondary defence mechanisms as those which operate during an encounter with a predator. The encounter may just involve the supposed or actual detection of a predator by the prey or it may involve an attack by a predator. Some secondary defences are entirely passive. The position spined sea urchin, diadema or the thick-skinned rhinoceros are adequately protected from attack. No behavioural specialisation is needed. Active forms of secondary defence include: (1) exaggerating primary defence, *e.g.* a camouflaged animal remaining motionless; (2) withdrawal to a safe retreat; (3) flight; (4) use of a display which deters attack; (5) feigning death; (6) behaviour which deflects the attack to the least vulnerable part of the body, to an inanimate object, or to other individual; and (7) retaliation.

Mostly, animals use both primary and secondary defence mechanism but very often different predators use different mechanism. In encounter with a predator, the prey animal may be able to use several alternative anti-predator behaviours. Decisions must be taken about, for example, whether to remain motionless, to flee, to display, or to fight. If one method is tried, without successful avoidance of attack, others may be tried. There are many possible strategies which involve alternative courses of action depending upon the characteristics and behaviour of the predator.

Secrecy

Many prey species protect themselves from predators by remaining in protected locations such as burrows, crevices, or huts. Cryptic colouration which functions to enable animals to blend in with the background, can be found in virtually all animal taxa. There are so many extreme examples found in insects which may resemble leaves, twigs or even bird droppings. Often a particular behavioural pattern is associated with protective colouration such that the animal orients itself in a particular relationship to its environment, usually remaining motionless.

Role of Camouflage

A phenomenon in which a species evolves to resemble its background is called camouflage. The nochid moth's defence is to seek escape in active flight. The opposite defence is to sit deed still and try to be invisible. Camouflage is of course an adaptation of appearance and colouration. The world is a patch-work of different colours: the animal is only camouflaged if it settles in the right place. Consider the European grasshopper Acrida turrita, it comes in green and yellow form. In nature the green form lives in green places and the yellow form in yellow and brown places. In a simple experiment, the German ethologist S. Ergene gave yellow and green grasshoppers and choice between yellow and green background. The green grasshooper fittingly tended to go and settle on the green backgrounds, and the yellow grasshoppers on the yellow.

Control over Colouration and Mimicry

There are some animals which are provided with poisonous or sickening substances in order to protect themselves from being eaten up by the predators. Some such animals make their own poisons; others takes them from sources in their environment. The wings of the monarch butterfly, for example contain powerful heart stopping poisons called cardiac glycosides. The monarch eats the poisons as a caterpillar, when its food plant is milkweed which contains cardiac glycosides. The caterpillar is not harmed by the poisons, it just stores them, and they are then retained by the adult.

For the behavioural problem of defence by poison we must turn from the prey to the predator. If the defence is the work, the predator must learn not to eat poisonous animals, because natural selection will not favour a trait, by which an animal, after it is dead, makes its attacker sick. If the trait is to evolve, it must ensure survival. The tactic used is to enable predators to learn to recognise sick-making prey, in which case they will avoid them. When a bird eats a monarch, as an adult or caterpillar, it will be violently sick within minutes, an experience it would learn not to repeat. The bird's problem then is to distinguish sickening from edible prey. Now, the colours of the monarch are bright gold, and ethologist suspect that the bright colouration evolved to make the monarch more memorable to birds. It is therefore called warning colouration. Experiments have shown that predators learn to avoid sickening prey, J.V.Z. Brower, offered the monarch butterfly as food to the Florida scrub jay. On the first meal of monarch the jays were violently sick, but after only a few trials they had learned not to eat monarchs but they continued to eat other, tasty food. What has not yet been conclusively demonstrated is that bird learn to avoid brightly coloured sickening prey more-quickly than equivalent duller coloured prey. Few ethologists would doubt, however, that such an experiment would be successful.

Means of Freezing

It is an another strategy of prey in response to predators. The presence of protective or cryptic colouration is often associated with this behaviour, as in the spotted white-tailed dur fawn (Odocoileus virgianus). Some animals carry this strategy further and feign death. For example the opossum (Didelphis virginiana), if harassed by a predator such as a dog, remains motionless on its back. The physiological correlation of this behaviour are not completely understood. Hognosed snake (Heterodon patyrhinos) exhibit similar behaviour, although they may precede it by threat behaviour in which the hognosed snake resembles a cobra. Freezing and feigning death may work because many predators are seen to respond only to the moving prey. For example, wolves will not attack prey that stand motionless (Mech 1970). Accompanied with this, rapid and agile locomotion provides the best and probably the most common means of escaping from predators.

Social Anti-predator Behaviour and its Characteristics

Confusion Effect: One of the main benefits of sociality is getting protection while living in groups. Living in flocks, herds or schools increases the chance of spotting predators. When approached or attacked by predators, groups usually become more compact. Fish schools, however usually form a vacuole around a predator. Tightly packed prey make a difficult target for the predator which must select an individual towards which it can aim its attack. The predator tries to isolate an individual from the group, any individual that strays from the group or is in any way different from the rest is likely to be attacked. For example, when experiments presented hawks with set of ten mice, the hawks preferred the oddly coloured mouse. (Muller 1971).

A primary means of protection is the confusion effect, described by Miller (1922). A variety of predators that attack swarming prey exhibit a lower prey capture rate, a longer period of hesitation, and more irrelevant behaviour than when they attack solitary prey. Factors that enhance the confusion effect are swarm number, swarm density, and the uniformity in appearance of swarm numbers, (Milinski 1979). The mechanism producing the confusion effect is not understood, but a predator seems to be more easily frightened when it feeds in a dense swarm of prey; perhaps the predator finds it harder to detect other animals that might prey on it. In laboratory experiments with three spined sticklebacks (Asterosteus aculeatus) feeding on water fleas (Daphnia), fish with experience in feeding on dense swarms fed more efficiently when tested with dense swarms than did those experienced only in feeding in low-density populations (Milinski 1979).

Detection: We might expect to find groups of mixed species if the combination of different sensory capabilities is advantageous. Baboons and ungulates are frequently together on the African plains. The arboreal habits and keen vision of baboons and the well developed olfactory system of ungulates probably increase the distance at which prey can be detected. It is clear that these different species recognise and respond to each other's alarm calls.

Social living increases the likelihood that predators will be detected and allows group members to spend more time in other activities. Individuals in large, dense colonies prairie drops (Cynomys spp.) spend less time in alert postures than do those in small colonies (Hoogland, 1979). Once predators are detected, alarm calls alert neighbours to danger. Marler (1955) has argued that in some birds the alarm call is difficult to locate because it is a high-pitched note in a narrow frequency range. The alarm-calling ground squirrels studied by Sherman (1977) emitted calls that were easily located, and calling squirrels were more likely to be caught than non-callers. Close relatives of the caller were likely to be close by and thus could benefit from this seemly altruistic behaviour.

Development of Anti-predator Behaviour: A baboon which is attacked by a leopard but which survives, learns something about the speed and attack methods of leopards. Future encounters with leopards will be different as a consequence of the experience. Field studies of the improvement in techniques of anti-predator behaviour are difficult, but many laboratory studies of avoidance conditioning mimic certain aspects of this type of behavioural change. Suppose that a rat in a cage learns to avoid shock by pressing a set of levers in a particular sequence when a certain combination of stimuli is presented. The rat may be using abilities, which would also be of use if faced with successive exposures to a predator. The effects of experience are often easiest to study in developing animals since some aspects of previous experience can be controlled.

In species where parental care is prolonged, the most effective anti-predator behaviour which young animals can show is often to recognise and maintain contact with the parents and the place where they are put by parents. The over-riding importance of finding the parent leads to some apparently anomalous responses to predators. Lorenz (1935) describes how newly hatched pre-costal birds, such as goslings, show no avoidance behaviour to a variety of moving objects including, for example, man, a goose predator.

A system which results in the young animal showing anti-predator behaviour to any novel change in its surroundings will operate when a real predator appears. Responsiveness to potential predators by young animals does not depend solely upon degree of novelty. It is also affected by the nature of the sensory input which results from the presence of the predator. Bodily damage, a looming object, or a loud noise will elicit more extreme anti-predator behaviour than the sight of small objects the sound of quieter noises, etc. (Hebb, 1946).

The existence of perceptual analysers in the brain which enable an individual to recognise dangerous events, for example, a looming object (Ewen, 1971). Some of these analysers are known to develop in a different way according to specific visual experience by others probably depend on general environmental factors such as light and metabolic stall at key times during embryological development. The average individual, therefore, would have some other predator detection mechanism whether or not there had been previous visual experience of the predator. An elaborate and specific possibility is a detector for flying bird of prey Lorenz (1939) and Goethe (1940) reported that young pre-costal birds did not respond to flying geese but showed a flight response when a hawk or falcon overhead.

Experiments by Schleidt (1961) showed that the results could have been due to habituation to the geese but not to hawks. Further work on any mallard duckling by Green, Oreen (1966) and by Mueller and Parker (1980), however, indicates that the hawk and goose silhouette can be distinguished and do elicit different reactions. The reactions were not clear anti-predator reactions in either case. The linking of the perceptual recognition to the command which initiates such reactions, therefore, may require experience which the experimental subjects of these experiments does not have.

4

The Osmoregulation

The maintenance of an optimal concentration of water and salts in the tissues and body fluids of animals is called *osmoregulation.* The term was first introduced by Hober (1902).

Isotonic, Hypotonic and Hypertonic Animals: Animals whose body fluids have the same concentration as that of the surrounding medium are called *Isotonic Animals.* Such animals never face the problem of osmoregulation. For example, most marine animals have their body fluids isotonic with sea water.

Animals which live in a medium of lower salt concentration are called *hypotonic animals.* In these animals, the water continuously enters into their bodies and dilutes their body fluids. Hence, they have to develop special mechanisms to get rid of the excess water that enters into their bodies. Examples of such animals are fresh water protozoans and crustaceous that expel their excess water through contractile vacuoles and excretory organs respectively.

Animals which live in a medium of higher salt concentration, *e.g.,* marine bony fishes, are called *hypertonic animals.* These animals face a danger of dehydration as the water is continually lost from their bodies. As such, they have to drink sea water and to evolve regulatory mechanisms by which excessive loss of water from their body is prevented.

Poikilosmotic and Homoiosmotic Animals: Aquatic animals which have entered media of different salinities may be either *poikilosmotic* or *homoiosmotic.*

Poikilosmotic Animals: Most of the aquatic invertebrate animals are evolved with tissues which tolerate dilutions of their body fluids and can adjust their internal medium to that of the external medium. Such osmotetbile or osmotically variable animals are called poikilosmotic.

Stenohaline: Poikilosmotic animals that can tolerate only slight changes in salinity are known as *stenohaline.* For example, the marine spider crab, *Maia,* when transferred to 80 per cent sea water (80 per cent sea water mixed with 20 per cent distilled water), puts up weight up to an certain extent due to quick entry of water into its body. Although the excretory organs of the crab work actively to eliminate the excess water in the form of a dilute urine, the crab loses large quantities of salts also along with the urine making the body fluid more dilute. Since the crab is not able to replenish the salts, it dies within 18 hours. *Arenicola* is another example of stenohaline condition.

Euryhaline: Poikilosmotic animals which can tolerate and withstand wide ranges in salinity are known as *euryhaline.* They are found in estuaries or river mouths or near sea shores where the salinity may fluctuate to a great extent. For example, the sea mussel, *Mytilus* can tolerate the dilution of sea water upto 4 per cent and lives unaffected. Sipunculid worms such as *Phascolosoma* ; *Aurelia* and *Aplysia* are also *euryhaline.*

Homoiosmotic Animals: These are *osmostable, i.e.,* they can regulate and maintain the osmotic concentration of their body fluid irrespective of the outside concentration.

Homoiosmotic animals living in fresh water adopt different methods for osmoregulation which are as follows:

(1) Presence of an impermeable cuticle.

(2) Restriction of surfaces to relatively small areas of respiratory and digestive systems.

(3) Development of excretory organs (nephridia or kidney) to produce hypo tonic urine.

(4) Possession of special cells (on the gills, in the gut or in the kidney) to absorb salts from the surrounding fluid to make good the salt loss through urine.

(5) Tolerance of dilute body fluids to minimise the expenditure of metabolic energy.

Mechanical Theory of Osmoregulation

I. Osmoregulation in Fresh Water Animals: Osm o-regulation in fresh water animals is effected by pumping out of excess water from their bodies. The salt loss through the excretion of water is made good by salt absorbing gills, skin, and various parts of the alimentary canal.

1. *Protozoa:* In fresh water protozoans such as *Amoeba, Euglena* and *Paramecium,* which live in a hypotonic medium, outside water enters into the body and dilutes the body fluid. This excess water is eliminated by the rhythmic pulsation of the contractile vacuole.

   Parasitic and urine protozoa do not posses a contractile vacuole.

2. *Crustacea:* The blood of fresh water crustaceans like *Palaemon* is hypertonic to the surrounding water. The latter, therefore, continuously diffuses into the blood through highly permeable gills. The excess water is excreted in urine by the antennary or green glands. Salt loss through urine is made good by the active uptake of chloride, sodium and potassium from the surrounding medium by the chloride cells of gills even if the concentration of salts in the medium is extremely low.

3. *Mollusca:* Fresh water mussels like *Unio* and *Anodonta* carry on osmoregulation through kidneys. Osmotic pressure of these animals is higher than the surrounding aqueous medium and as such water diffuses into the tissues. This water is eliminated by the kidneys during excretion.

4. *Fresh Water Teleosts (Bony Fishes):* These are *hyperosmotic* to the surrounding medium, *i.e.,* their blood is more concentrated than the external medium. As a result, outside water enters through the thin permeable skin, the smooth surface of the gills and lining of oral and pharyngeal cavities. This water is quickly transported to the well-developed glomerular kidneys which reabsorb desirable solutes and secrete a dilute urine. Thus the blood concentration is restored. However, alongwith the urine, some salts also pass out. The salt loss is made good by chloride cells present on the gill epithelium. These cells are capable of absorbing the small quantities of salts present in the fresh water. In this way, kidney of fresh water fishes serves primarily as filter and reabsorber of solutes.

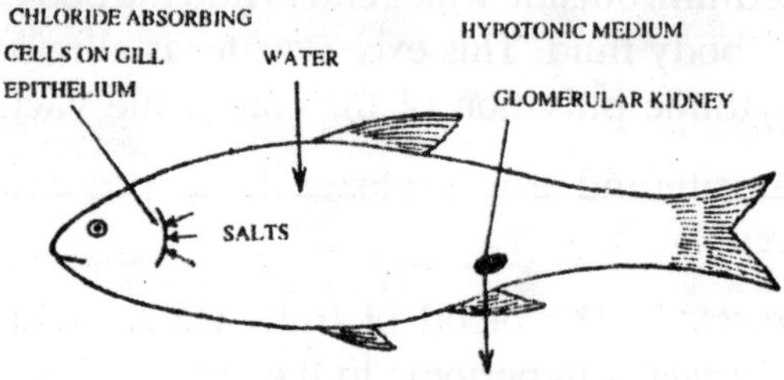

Osmoregulation in a fresh water teleostean fish.

5. *Fresh Water Elasmobranchs (Cartilaginous Fishes): Pristis* (sawfish) and *Charachariasgiganticus* (Shark), which often enter the large rivers and other fresh water habitats, have a tendency to retain some urea in their blood and tissues. As a result, large quantity of water

enters the body by osmosis. This water is eliminated by the large glomerular kidneys.

6. *Amphibia:* Fresh water which contains a small percentage of salts diffuses into the body through smooth skin of amphibians. The well developed glomerular kidneys serve to remove the excess water which is excreted as dilute urine. The salts lost through the urine are replenished by reabsorption of ions from the urine.

Osmoregulation in Marine Animals

1. *Hagfish:* A few marine animals such as the hagfish, *Myxine,* maintain body fluids about as salty as the sea waters of the surrounding ocean (isosmotic) and so do not tend to lose water by osmosis. However, the glomerular kidney of the hagfish is an ion-regulating structure; it removes Ca^{++}, Mg^{++} and SO_4^- and reabsorbs K^+ and Cl^-.

2. *Marine Teleosts:* In the marine bony fishes, e.g., *Opsanus* and *Lophius,* the body fluids are hypotonic to the marine environment, having a solute concentration only about 1/3 that of sea water. Consequently, they lose water through the smooth surfaces of the body like gills and oral epithelia. Thus they are constantly in danger of losing so much water to their environment that the solutes in their body fluids may become extremely concentrated and the fish may die of osmotic desiccation. Therefore, to compensate for their osmotic water loss, they *drink* sea water. This restores their water content but leads to a new problem—how to eliminate the excess salt ingested. This problem has been solved by the evolution of special "chloride secreting cells" in the gills that excrete excess salt. Hence these fish can take in salt water and still remain hypotonic. Water loss through urine is minimised by the development of *aglomerular* kidneys. Further, the

distal convoluted tubule of the nephron is also absent. Consequently, urine flow in *Opsanus* is only 2.5 ml/kg per day as compared to 300 ml/kg per day in a fresh-water teleost.

3. *Marine Elasmobranchs:* Marine cartilaginous fish such as *Scoliodon* and *Stegostoma* have body fluids isotonic with sea water. They achieve their isotonicity in a different way. Because they have an unusual tolerance for urea, instead of constantly, pumping this waste out, as do all other fish, sharks retain a higher concentration of it in their blood equal to that of sea-water. Though the osmotic imbalance is removed in this way, the ionic concentration of the body fluids and the surrounding sea-water remains different. The marine elasmobranchs achieve their ionic regulations by actively removing the excess salts with the help of a *rectal gland.* They do not have chloride secreting cells in their gills.

4. *Migratory Fishes:* Catadromous fishes like the eel, *Anguilla bengalensis* migrate from fresh-water to sea, while anadramous fishes like Salmon, *Salmofario,* migrate from the sea to fresh-water for spawning. Such fishes have to regulate their blood tonicity in the two contrasting media. It has been observed that the blood concentration of eel is hypertonic when it lives in the fresh-water and it shows the same pattern of osmoregulation as fresh-water teleosts. On the other hand, blood concentration of the spawning eel in the marine medium is hypotonic to the sea-water. Therefore, it maintains itself against an osmotic desiccation (a) by drinking sea-water and *(b)* eliminating the excess salts by the chloride secreting cells of the gills.

5. *Marine Reptiles:* Some reptiles like turtles, *sea-snakes* and the *iguana* are able to maintain the osmotic and ionic concentration of their body fluids by the same mechanisms as met within marine teleosts. They drink

sea-water and consume salty prey. Thus they are able to absorb the water but eliminate the salt through special "salt glands".

Turtles have special glands in their heads (lachrymal glands and pre-maxillary glands) that can excrete sodium chloride at a concentration about twice that of sea water. Since ancient times, turtle watchers have reported that these great armoured reptiles come ashore, with tears in their eyes, to lay their eggs, but it is only recently that biologists have learned that this is not caused by an excess of sentiment — as is the case with Lewis Carroll's mock turtle—but is, rather, a useful solution to the problem of excess salt from ingestion of sea water.

6. *Marine Birds:* Sea birds such as *petrels, cormorants, penguins* and *herring gulls* eliminate their excess salts, chiefly sodium chloride, in the form of a salty fluid through their nostrils. The salt glands are located in their nasal chambers. In the herring gull, the concentration of sodium chloride thus got rid off is about five times its concentration in the blood.

7. *Marine Mammals:* Whales, dolphins, porpoises and seals do not drink water, hence they depend upon food for their water requirements. These mammals never sweat and in this way avoid loss of water by perspiration from their skin.

Osmoregulation in Terrestrial Animals: Since terrestrial animals do not always have automatic access to either fresh or salt water, they must regulate water content in other ways, balancing off gains and losses. Terrestrial animals gain water by drinking fluids, by eating water-containing foods, and as an end product of the oxidative processes that take place in the mitochondria. When 1 gm of glucose is oxidised, 0.6 gm of water is formed. When 1 gm of protein is oxidised, only about 0.3 gm of water is produced.

Oxidation of 1 gm of fat, however, produces 1.1 gm of water because of the high hydrogen content of fat (the extra oxygen comes from the air).

The main problem of terrestrial animals is the loss of body water content. It is solved by a combination of some of the following adaptations:

1. Presence of a highly impermeable cuticle (arthropods) or skin covered with scales (reptiles), feathers (birds) or fur (mammals) and deposited with wax or keratin. Such a water proof cover prevents loss of water by evaporation.
2. Elimination of nitrogenous wastes, as urea or uric acid with little or no loss of water. In reptiles and birds, loss of water is restricted due to uricotelism. They excrete a semi-solid urine in the form of uric acid crystals. Reabsorption of water from faeces by the rectal glands and cloacal wall further minimises water loss. In mammals and in some birds there is a water absorbing column called Henle's loop in the uriniferous tubule of the kidney which minimises the water loss through urine.
3. Ability of water absorption through the body surface. The skin of the desert lizard, *Moloch horridus* absorbs water like a blotting paper when the atmosphere is humid or the animal is drenched in rain.
4. Production of metabolic water by the oxidation of food stuffs, particularly the fat. Some animals living in extremely dry habitats depend upon metabolic water for their vital activities. The Kangaroo rat, Dipodomys, of the American desert for example, may spend its entire life without drinking water. It is not surprising that it prefers a diet of fatty seeds, which yield a large amount of water on oxidation. Analysis shows that the Kangaroo rat is highly conservative in its water expenditures. It has no sweat glands, and being

nocturnal, it searches for food only when the external temperature is relatively cool. Its faeces have a very low water content, and its urine is highly concentrated. Its major water loss is through respiration, and even this loss is reduced by the animal's long nose in which some cooling of the expired air takes place, with condensation of water from it.

If the Kangaroo rat is fed high-protein seeds, such as soybeans—the oxidation of which produces a large amount of nitrogenous waste and a relatively small amount of water—it will die of dehydration unless some other source of water available.

On the average, a human takes in about 2,300 millilitres of water (about 2.5 quarts) a day in food and drink, and gains an additional 200 millilitres a day by oxidation of nutrient molecules.

Water is lost from the lungs in the form of moist exhaled air, is eliminated in the faeces, is lost by evaporation from the skin, and is removed from the blood and excreted as urine. The latter is usually the major route of water loss. In a normal human adult, the rate of water excretion in urine averages 1,500 millilitres a day, but the actual amount produced may vary from 500 to 2,300 millilitres.

The *cleidoic eggs* of reptiles and birds are provided with a large amount of yolk and enclosed by a hard shell. The eggs are laid on land. The developing embryo has limited supply of water, chiefly *metabolic* in nature. The nitrogenous wastes are converted into non-toxic uric acid crystals which are stored in an *allantoic sac* to be excreted at a later stage.

5. *Tolerance of Dehydration and Fluctuation in Internal Temperature:* The *camel* "ship of the desert", can tolerate the loss of more than 25 per cent of their body weight in water, going without drinking for as long as an entire week in the summer months, three weeks in the winter. Besides, the camel can tolerate a fluctuation in

internal temperature of 5° to 6°C. It is estimated that the camel saves as much as 5 litres of water a day as a result of these internal temperature fluctuations. Its body temperature can vary from 34° to 40.5° C. Without taxing its regulatory machinery like sweat glands (Man cannot tolerate a temperature fluctuation of 1°C). The camel never sweats till the body temperature rises to 41° C. Therefore, much water loss is saved by this method. Moreover, the camel excretes a much more concentrated urine, in other words, it does not need to use so much water to dissolve its waste products.

6. Behavioural adaptations tending to confine the animals to regions of high humidity and to burrow into the soil.
7. Restriction of activity to those periods when the temperature is not severe.
8. Regulatory mechanisms to check water loss (diuresis) through nervous and chemical coordination (secretion of anti-diuretic hormone, ADH, from the posterior lobe of the pituitary gland and Aldosterone from the adrenal cortex). ADH increases the reabsorption of water from kidney tubules whereas *aldosterone* is concerned with ionic balance. In mammals, aldosterone increases the reabsorption of sodium and promotes the renal excretion of potassium.

Role of Osmoregulation

With the help of their several structures all the living organisms maintain a proper amount of water and salts in their bodies. This regulation of the water and salt contents of the body is termed as *Osmoregulation.* However, the *osmotic regulation* and *ionic regulation* are sometimes treated as the different mechanisms. But in most animals the osmoregulatory mechanism includes both of them because the regulation of one (water) always affects the other (salts).

Marine, Brackishwater, and Freshwater Fauna: It is

generally considered that life originated in the sea. Part of the evidence for this is the extreme variety and richness of the present-day marine fauna and there is marked impoverishment as one proceeds into brackishwater and freshwater.

Many marine organisms are stenoheline, highly intolerant to dilution of the external medium, and some groups (Cephalopoda, Echinodermata) are virtually absent from brackishwater and freshwater. But from stenohaline marine fauna many forms have managed to adapt to and colonise brackish and ultimately fresh water. The abundance of freshwater insects and branchiopod Crustacea and their rarity in the sea is noteworthy.

There is also evidence that some groups have secondarily returned from freshwater to the sea, particularly fish groups—elasmobranchs, teleosts, and coelacanths. Thus, movement from sea to freshwater and *vice versa* has been possible through osmotic and ionic regulation. The successful penetration of marine animals into fresh water required the conquest of an osmotic problem which was the opposite of any which might exist in the sea. Thus, osmoregulatory devices have been developed in accordance with the evolutionary trends of animals.

The concentration of sea water and brackish water can be defined by the salinity or be given directly in terms of the concentration of Na^+ and Cl— ions. The concentration of open sea water varies from 30 to 40 per cent O. Water more dilute than this would be regarded as brackish water. Usually a salinity < 0.5 per cent O (7.3 mM Cl^-) is regarded as freshwater.

Isotonic, Hypotonic and Hypertonic Animals: Animals whose body fluids have the same concentration as that of their surrounding medium *(isotonic* or *isosmotic animals)* never face the problem of osmoregulation as long as they live in such a medium.. For example, most marine animals are isotonic, as they have their body fluids isotonic with sea water. Animals which live in a medium of lower salt concentration *(hypotonic* or *hyposmotic animals)* have to face the difficulty of hydration and hence they have evolved special mechanisms to get rid of the excess water that usually enters their body and dilutes their body fluids.

For example, fresh water protozoans and crustaceans are hypotonic and eliminate the excess water by their contractile vacuoles and excretory organs respectively. In a similar way animals which live in a medium of higher salt concentration *(hypertonic* or *hyperosmotic animals)* have to evolve regulatory mechanisms by which excessive loss of water from the body is prevented. Since these animals *(e.g.,* bony fishes of the sea) are in constant danger of dehydration, they compensate for the water loss by drinking the surrounding water or by other means.

Process of Osmosis and Osmotic Pressure

In order to understand the physiological reactions taking place in the mechanism of osmoregulation, we must have some knowledge of the behaviour of substances in solution. When a weak and a strong solution are separated by a semi-permeable membrane, there is always a transfer of the solvent from the weaker solution to the stronger one. This process of selective transmission of a liquid in preference to another or a solvent in preference to the solute through a semi-permeable membrane is termed *osmosis.* This process continues until the hydrostatic pressure due to the flow of the liquid or solvent has attained a value sufficient to stop further flow.

The excess pressure which is just sufficient to stop the flow through the membrane is called the *osmotic pressure* of the stronger solution. The osmotic pressure is proportional to the concentration of the solution, or in other words, the greater the concentration of a solution the greater would be the osmotic pressure of it. Thus, the magnitude of the osmotic pressure or the *osmotic concentration* depends on the number of particles dissolved per volume of the solvent and not on their size or nature. If several different substances are dissolved, the osmotic concentration depends on the sum of all the dissolved particles.

One solution is said to be *isosmotic* with another when they have equal osmotic concentration. Similarly a *hyposmotic* solution is more dilute and a *hyperosmotic* one more concentrated than a comparison solution. The osmotic pressure of a dilute solution is given by the Van't Hoff Equation:

$$\pi = RTC$$

where π = the osmotic pressure

C = the total concentration of non-electrolytes and ions present

R = gas constant

T = temperature (in Kelvin)

Dimensionally, it is correct to express C as moles and g-ions per litre of solvent. However, it is more usual to express concentration as moles and g-ions per kg of solvent *(molal scale)*. This is numerically correct in above equation for an *aqueous* solution. This total concentration is sometimes, incorrectly, termed *osmolality*.

It should be noted that above equation is only approximate in the upper range of concentration found in many biological situations. However, if the osmotic pressure of a solution is known, the ratio π /RT defines a value C that is *osmolality*. Osmotic pressure is frequently expressed using this *osmolal scale*. It is difficult to measure the osmotic pressure of a solution directly, but other properties related to osmotic pressure can be more easily determined. These properties, called the *colligative properties*, are all determined by *water activity* in the system. Van't Hoff equation shows that osmotic pressure of a solution at any given temperature T depends only upon the molar concentration C, *i.e.*, on the number of moles (and hence on the number of molecules) present per unit volume, quite independent of the nature of the solute. Thus, osmotic pressure of a solution is a colligative property.

Animal Response to Osmotic Conditions of the Medium: There are two extreme patterns of response to osmotic conditions of the environment. Animals may be osmotically labile (dependent) and their body fluid concentration may change with the medium. These are *osmoconformers.* Other animals are osmotically stable (independent) and when the medium changes, the internal concentration of the body fluid remains constant or unchanged. These are *osmoregulators.* The terms *poikilosmotic* and *homoiosmotic* are often applied respectively to osmoconforming and osmoregulating animals. Osmoconformers can tolerate wider variation in their internal osmotic concentration, while osmoregulators can withstand a wider environment range. However, there are all gradation between the extremes of liability and constancy. An animal may conform osmotically in one concentration range and regulate in other.

Euryhaline and Stenohaline Animals: Poikilosmotic animals which can tolerate wide ranges in salinity are known as *euryhaline.* For example, *Mytilus* when left in sea water diluted to a great extent, can adjust the salinity of its haemolymph to exactly the same concentration and lives unaffected. Similarly *Aplysia,* left in 75 per cent sea water increases in weight by 20 per cent in 2 hours, but loses salts, and water in the next 4 hours. In this way its original volume is regained and it is able to live with this diluted blood in the changed water content.

On the other hand poikilosmotic animals which cannot tolerate wide ranges in salinity are known as *stenohaline.* For example, a marine crustacean, spider-crab *(Maia),* if left in 80 per cent sea water increases in its weight up to a certain extent due to the quick entry of water into its body. The excretory organs of the animal activate to get rid of the excess water in the form of dilute urine. However, along with the excreted water some salts also escape out, making the body fluid more dilute, and as a consequence the animal dies in 2 to 4 hours.

Osmoregulation in Freshwater Animals: All freshwater invertebrates, as well as fishes and amphibians, are hypertonic (hyperosmotic) to the surrounding water. Since they are never completely impermeable, they have steady inflow of water and some loss of salt. They eliminate the excess amount of water as urine, and by elaborating a urine of very low concentration, they keep salt loss in the kidney to a minimum. However, some salt is lost and must be made up by food or by active absorption from the very dilute water. Numerous freshwater invertebrates can absorb salt actively and probably all of them have some salt absorbing mechanism.

Freshwater protozoa living in *hypotonic* medium, eliminate excess water by their contractile vacuoles. The pulsation rate of the vacuole increases when more water penetrates into the body. In freshwater crustaceans the excess water is eliminated by their excretory organs, the antennary or green glands.

The small amount of salt lost with the dilute urine is restored by active absorption of salts by the gills from the freshwater. Freshwater molluscs, *e.g., Unio* and *Anodonta* also excrete a large amount of dilute urine and absorb salts to make up the loss. Similarly, in freshwater fishes, the water that enters the body is eliminated by the kidney. The urine is quite low in salt and is highly hypotonic, as it is in most freshwater animals. The salts loss is balanced by active absorption of salt in the gills.

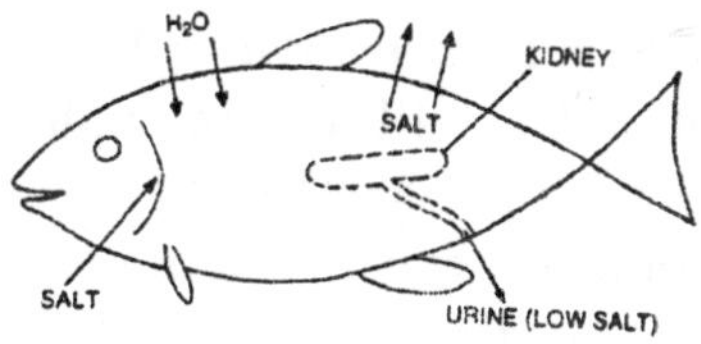

Osmoregulation in a freshwater bony fish.

Aquatic insects vary considerably in the permeability of the cuticle to water. Gills are more permeable than body surfaces, and larvae and nymphs tend to be more permeable than adults. Aquatic insects in air tend to lose more water by evaporation than do terrestrial insects. Larvae of aquatic insects conserve or extract salts not with their gills but with their guts. Either the rectal cuticle or special anal papillae developed by some larvae, such as *Aedes aegypti,* can absorb salts and the gut wall is histologically differentiated in these areas. In *A. aegypti,* the anal papillae are responsible for about 90 per cent of salt conservation.

The fresh water crayfish, *Astacus,* secretes urine equal to about 4 per cent of its body weight in 24hr, the osmotic concentration of which is about one-fifth that of its blood. The antennary organ of this animal has a long canal between labyrinth and bladder.

The canal functions as a water-secreting segment. Salt loss is compensated for by the active uptake of salts from the medium, even when the external salt concentration is very low, Cl^-, Na^+ and K^+ all being absorbed by the gills. *Astacus* can extract Na^+ from water containing as little as 0.02 to 0.09 mM/litre, and Cl^- from a medium hardly less dilute.

Amphibians are very similar to freshwater fish, although the adult animals have no gills and the absorption of salts takes place from the whole skin surface. The common frog can extract Cl^- from a medium containing only 0.01 mM/litre. As they have a low salt concentration, they have not been able to adapt to a marine life, for in salt water they would become dehydrated and salt would enter the body.

Osmoregulation in Brackish Water Animals: Brackishwater is generally defined as dilute sea water that ranges in concentration from 1.5 per cent to 90 per cent of that of pure sea water. Most brackish waters contain both marine animals which can tolerate low salinity, and true brackishwater animals which are not found in the sea. Typical marine animals in brackish water include the shore crab *Carcinus maenas* and the mussel *Mytilus edulis.*

These marine forms tend to disappear from the brackish water fauna when the water becomes equivalent to about 15 per cent sea water concentration. An example of a true brackish water animal is the prawn *Palaeomonetes varians,* which can live in water ranging from 0.5 per cent to 100 per cent sea water, but does not in fact live in genuinely marine habitats.

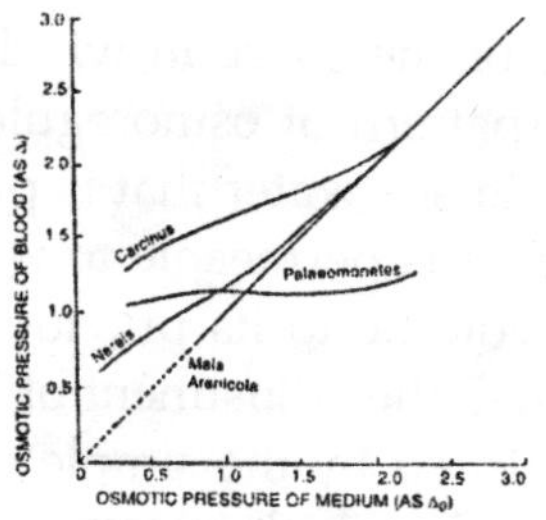

Osmotic pressures of the body fluids of some marine or brackish water animals plotted against the osmctic pressure of external medium.

When we plot the osmotic pressures of the body fluids of a number of brackish water or marine animals against the osmotic pressure of the external medium, different types of curves are obtained. The greater the deviation of the curve from a linear curve that passes back through zero if it is extrapolated back, the greater is the ability of the animal concerned to regulate its blood concentration despite the osmotic effects of the diluted medium. Linear curves are characteristic of the spider crab *Maia* and the lung-worm *Arenicola marina.* In *Nereis* and *Carcinus,* the osmotic pressure of the blood is at first linear, but as the dilution of the medium is progressively increased the internal osmotic pressure follows the external osmotic pressure to a decreasing extent, the blood becoming increasingly hyperosmotic to the medium.

The ability to regulate is greater in *Carcinus* than *Nereis diversicolor* and therefore it is no wonder that *Carcinus* can penetrate further into brackish water than *Nereis.* The curve given by *Palaeomonetes* is quite different from the others, reflecting the true brackish water nature of this prawn. Even in 100 per cent sea water its blood concentration is equivalent to only 70 per cent sea water. But it can maintain a blood concentration equal to 50 per cent sea water when the external medium is diluted to 5 per cent sea water. Below this level some further dilution of the blood occurs, and the animal dies in media more dilute than 0.5 per cent sea water.

In the planarian *Procerodes ulvae,* which lives in tidal

estuaries and areas of the shore in which fresh water enters the sea, a different pattern of osmoregulation is exhibited. If *Procerodes* is placed in sea water that is progressively diluted, the animal swells as it is permeable to water. But eventually, it adjusts its body volume to its previous level. By chemical analysis of the animal, the adjustment of its body volume can be shown to be achieved by the simple expedient of actively secreting ions outwards, which is done to a striking extent. In fresh water, *Procerodes* possesses an internal salt concentration that is only one-tenth of that in sea water. The limitations of such a method are obvious, and the fact that the blood concentration reached in fresh water is dangerously low is illustrated by the observation that *Procerodes* cannot live indefinitely in fresh water. Such a method of adjustment is only suitable for an animal living in tidal conditions.

Osmoregulation in Marine Animals

Most invertebrates living in the sea have an osmotic concentration in their body fluids which equals that of the surrounding seawater but they can regulate their ionic composition. When a marine invertebrate is placed in dilute sea water, the water enters the animal due to the higher osmotic concentration in the animal. But after sometime a state of osmotic equilibrium is attained with the new environment. Most animals that are exclusively marine can tolerate only a limited change in the concentration of their medium. If the medium becomes too dilute, they would die.

The shore carb, *Carcinus,* can actively resist changes in its osmotic concentration because it has the power of osmoregulation. *Carcinus* can live in brackish water. In normal sea water it is in osmotic equilibrium, but in dilute sea water it maintains a higher osmotic concentration in the blood than that in the surrounding water. In this situation, the internal osmotic concentration in the crab causes a steady inflow of water, which must be eliminated again.

Since the crab has a kidney, the so-called green gland or

antennary gland, the water can be excreted. But curious fact is that the urine formed is not dilute and has the same osmotic concentration as the blood. A high salt concentration in the crab is, therefore, maintained by the gill's ability to secrete salt into the blood. The salt naturally tends to diffuse from the concentrated blood to the more dilute water. For example, in 40 per cent sea water *Carcinus* loses 4.4 mM/Kg body weight/hr of sodium ions through diffusion from its permeable surfaces, and 3.8 mM/ kg/hr in its urine. This total net loss of 8.2 mM/kg/hr is recouped entirely through the active uptake of Na^+ ions from the medium against the concentration gradient.

The ability of a marine animals to penetrate into a brackish or estuarine environment depends on three different abilities, or a combination of them. These are:

(i) The animal may temporarily exclude dilute sea water, for example, a clam or oyster may close its shells when the changing tides bring highly diluted sea water ;

(ii) The animal may tolerate the decrease occurring in the concentration of its body fluids as the salinity of the water decreases temporarily;

(iii) The animal may actively take up salt from the water and maintain a higher osmotic concentration than the surrounding water. This form of osmotic regulation will enable the efficient animal to live in dilute water, and may even permit him to penetrate into freshwater. The Chinese mitten crab, *Eriocheir*, can live in the rivers of Central Europe, but it must return to the sea to breed, for the eggs do not tolerate fresh water. This crab, then shows the transition in osmoregulation that is necessary for life in freshwater.

Marine animals placed in more concentrated sea water will usually show increased blood concentrations equal to that of the water. But some including the mitten crab, can maintain concentrations lower than the surrounding water by actively transporting salt from the blood to the water. Such regulation becomes fairly important in the animals living in salt lakes, such as the fairy shrimp, *Artemia.* It can exist in salt solutions containing as little as 0.26 per cent NaCl and in concentrated brine containing as much as 30 per cent salt. The osmotic concentration of the blood is relatively independent of the medium and increases only slightly as the salt content of the water increases.

The myxinoids (hagfishes) are exclusively marine. They are osmotic with the medium and show considerable ionic regulation by varying the total osmoconcentration largely by charges in amino acids. Lampreys show both hyposmotic and hyperosmotic regulation. When marine lampreys *(Petrotmyzon marinus)* enter freshwater they stop drinking, but if returned to brackishwater they drink and extrarenal extraction of Na^+ and Cl^- is activated. In freshwater, they can actively absorb chloride.

The *bony fishes* in the sea are not in osmotic equilibrium with their *hyperosmotic* medium, for their osmotic concentration is about one-third that in sea water. Since they are not impermeable, they lose water to the more concentrated sea water and are actually in constant danger of dehydration. They compensate for the water loss by drinking the very sea water and by absorbing water and salt (Na^+ and Cl^-) from the intestine. The fish kidney is unable to produce a concentrated urine to help eliminate the salt. Instead, the gills secrete the excess salt ions (Na^+, K^+ and Cl^-) into the sea water. Since this secretion is from a lower concentration in the blood to a higher one outside, it is an active transport and needs energy. Some ions are lost in the faeces and urine (Mg, SO_4, Ca, K and PO_4).

The *elasmobranchs* (cartilaginous) fishes have solved their osmotic problem with the help of their kidney. Their blood contains sufficient amounts of urea and trimethylamine oxide, thereby maintaining an osmotic concentration equal or more to that of the sea water. Thus, for osmoregulation, their kidney possesses a special mechanism for pumping urea back into the blood almost as fast as it passes through the glomerular filter.

Even the young elasmobranch is provided with its store of urea, either directly from the parent as in the viviparous dogfish *Squalus*, or by its being placed in an egg case impermeable to urea. By the time they hatch or are born, the young animals have developed their own urea conserving mechanism.

As the osmotic concentration of the blood is often 10 to 50 mOsm/kg more than that of the sea water, marine elasmobranchs can, therefore, use mechanisms similar to those of freshwater fish for the excretion of water. There is a tendency for osmotic inflow and excess water is removed by the kidneys. However, there is also an inward diffusion of salt which must be eliminated. Excess salt is excreted not by the gills but by specialised organs, the *rectal glands*. For example, the dogfish *Squalus acanthias* secretes about 2 ml/kg/hour of solution through its rectal glands; this is a volume about equal to the urine output of the animal. The rectal glands are responsible for eliminating excess monovalent ions, whereas the kidneys eliminate divalent ions. The gut plays little or no role in water or solute balance in elasmorbranchs. Some urea which diffuses through the gut wall is eliminated in the faeces.

The maintenance of a high level of urea is clearly an adaptation to marine conditions, and is presumed by most authorities to indicate that elasmobranchs too were originally a fresh water group, but that all returned to the sea at some stage and were enabled to live there more easily through the development of urea conservation. On this view, fresh water elasmobranchs have returned once more to their habitat of origin. Since they possess a blood urea concentration only little below that of their marine relatives, it seems that they have been able to reverse this evolutionary change to only a limited extent. The presence of this amount of urea in the blood is additional evidence for a return to fresh water from the sea, since it would not have arisen in fresh water.

Marine turtles, marine iguanas, and sea snakes all have plasma osmotic and ionic concentration similar to those of bony fishes. Their interguments are highly impermeable, and they also lack the exposed permeable surfaces of gills because they are air breathers. Their major problem is a continuing water loss because they must produce sufficient urine to eliminate metabolic wastes and they must lose some water in respiration. However, they are not known to drink water and the physiological mechanisms used to control water are not known. The urine is produced at a low rate, and the kidneys of these animals have a low glomerular filtration rate. The salt intake occurs, in part at least through the food (*Schmidt-Nielsen* and *Davis*, 1968).

In marine birds such as sea gulls, cormorants, and pelicans that drink sea water, the excess salt is eliminated as NaCl by nasal glands of their heads. Marine mammals do not drink water to any great extent and rely on their food as the sole source of water. Whales are a good example. They have a slightly higher blood osmotic concentration than terrestrial mammals and this also aids in osmoregulatory problems.

Most marine mammals feed on teleosts rather than marine invertebrates. This is an advantage because teleosts have already performed the osmotic work of eliminating excess salts from their tissues. Whether marine mammals possess mechanisms for extrarenal salt loss is not known, but they can produce a hyperosmotic urine when stressed.

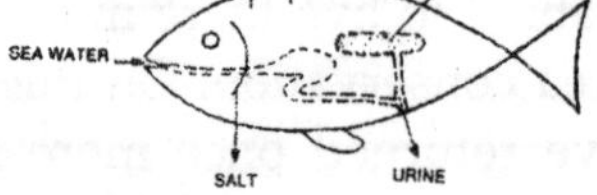

Osmoregulation in a marine bony fish.

Osmosis in Migratory Fish

When an eel (*Anguilla*) reaches the freshwater habitat on its upstream migration, it faces problem of salt depletion and over-hydration as compared with the tendency towards dehydration and salt excess in the ocean. A young salmon (*Oncorhynchus*) migrating downstream to its ocean habitat comes up against these conditions in reverse. It follows then, that both *anadromous* (marine fishes that migrate from sea to rivers for spawning, *e.g.*, Salmon) and *catadromous* (freshwater fishes that migrate to sea for spawning, *e.g., Anguilla)* fishes must be versatile in their osmotic adjustment, have glomerular kidneys which can adjust to differences in urine volumes due to different salinities and possess gills and oral membranes capable to cope both with the uptake and the secretion of certain ions against existing diffusion gradients.

Experiments with eel have suggested that the chloride cells can function both as secreting and absorbing units. However, it should be noted that *diadromons fishes* are adjusted either to the freshwater or to the saltwater phase of their life cycles and that such adjustments are primarily dependent on genetically determined physiological changes.

The sea lamprey *(Petromyzon),* for example, after having entered freshwater in order to spawn cannot adjust to salinities higher than 17 parts per thousand, and young salmons *(Oncorhynchns, Salmo salar)* cannot successfully enter the sea before their chloride cells are well developed. The case of transition depends on anatomical peculiarities such as the gill to body surface ratio, well illustrated in different races of *Gasterosteus,* which show differential rates of adjustments to freshwater on their spawning migrations along the coasts of western Europe.

Changes in the endocrine activity are usually simultaneous with, or precede, changes in salts and water balance mechanism. The pituitary, the thyroid, and the gonads are primarily concerned with changes in physiological adjustment prior to and during migration. An increase in thyroid activity has been reported for salmon on downstream migration, possibly to facilitate the energy requiring process of salt excretion in sea water.

Osmoregulation in Terrestrial Animals: All terrestrial animals lose water to their environment and therefore preserve water to get rid of the danger of dehydration. The only truly successful land animals are higher arthropods and higher vertebrates. All other animals depend more or less on a moist environment. If an animal has to maintain its normal water content, all the water it loses must be exactly made up by the water it gains. If the total gain in water is equal to the loss, then the animal is said to be in *water balance;* and if the water loss is not covered by an equal gain, the animal is said to be in *negative water balance.* This situation is dangerous and is avoided by the animals.

There are several ways by which water is gained or lost in land animals. The water gain includes drinking water, water in food, and the oxidation water formed inside the body in the combustion of food materials, while water loss occurs through evaporation from the lungs and skin, urine formation, and faeces.

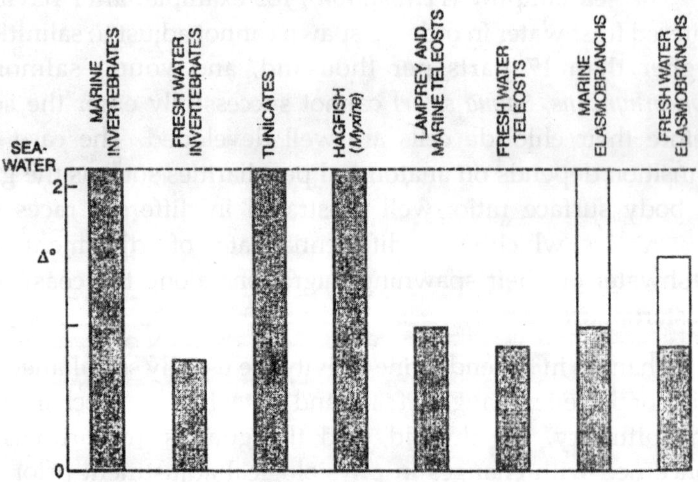

Comparative histogram of the osmotic pressures of the body fluids of invertebrates and fishes. The proportion due to urea is represents by clear areas.

To prevent the loss of water from the body, the land arthropods (arcahnids, myriapods and insects) have a chitinous coat. The insects have been the most successful in minimising surface evaporation. They have a thin waxy cuticle of such unique properties that it not only prevents evaporation but can also take up water from a moist environment.

A unique property of the cuticle of insects and ticks is the capacity to absorb water from an atmosphere which is humid but less than saturated. Water absorption from the atmosphere has been demonstrated in ticks, mites, fleas, *Tenebrio* larvae, and lepidoptera. Another adaptive mechanism is behavioural selection of "preferred" humidity. Slightly desiccated animals tend to go to higher humidities and hydrated ones to drier air. For example, adult *Drosophila* from a normal culture prefer drier air, but they reverse this preference if their water content is reduced by 12 to 16 per cent.

In some lepidopteran caterpillars, the third antennal segment has three receptor neurons which increase their discharge with a rise in temperature, and three others which increase spontaneous firing in moist air and decrease it in dry air. Such humidity preferences play a role in the distribution of many insects.

Many terrestrials insects excrete nitrogenous wastes in solid or semi-solid form and reabsorb water in the posterior part of digestive tract. In a hydrated *Periplaneta* the rectal fluid (urine) is hyposmotic to haemolymph; in dehydrated *Periplaneta* urine osmotic concentration may be twice that of haemolymph (*Wall* and *Oschman*, 1970). In the desert locust *Schistocera,* the rectal fluid is always hyperosmotic and only a small part of the solute is inorganic ions. In the mealworm *Tenebrio molitor,* faeces enter therectum as liquid and leave the anus as dried pellets. The animal has perirectal malpighian tubules, which can take K^+ from the haemolymph/ and transport out of rectum through an osmotic gradient.

Of all the invertebrates, only insects can produce markedly hyperosmotic urine and are able to conserve water. However, the mechanisms by which they accomplish this are quite different from those in mammals. In terrestrial insects, K^+ is cycled and water can be moved via osmotic gradients in restricted intercellular channels back to the haemolymph, to leave hyperosmotic excreta.

In other higher vertebrates, protection against surface drying is afforded by horny scales (reptiles), feathers (birds), and hair (mammals) of the skin. Another means of water conservation in birds and mammals is that their kidney excretes a concentrated urine. This ability of their kidney depends upon the *loop of Henle,* the function of which is the resorption of further water. Those animals *(e.g.,* desert rat) which can produce the most concentrated urine have the longest loops of Henle.

In mammals, the volume of urine is controlled by the anti-diuretic hormone, (ADH), that is secreted by the posterior lobe of the pituitary gland. The hormone plays an important role in regulating water reabsorption by the renal tubules. An under-secretion of ADH causes a disease in which a huge amount of urine is excreted (diabetes insipidus). The osmoregulatory mechanisms of terrestrial animals, thus, operate by reducing the loss their water content to minimum and involve several organ systems, including the endocrine system. However, the lack of such protection keeps some groups confined to a moist environment, *e.g.,* Onychophora and Amphibia.

Hormonal Control of Osmoregulation: Endocrine secretions control osmotic and ionic balance in both invertebrates and vertebrates chiefly by action on gills, skin and excretory organs, and on tissues in relation to extracellular ions. The role of following hormones is considered important.

Neurohypophyseal Hormones: The neurohypophyseal polypeptides are best known for their function in water and salt balance of vertebrates. These are produced by the supraoptic nucleus of the hypothalamus, pass along the infundibular stalk to the neurohypophysis, where they are stored and released on demand. Those with a basic amino acid such as arginine or lysine in the 8-position are called *vasopressins* and have marked anti-diuretic (ADH) or water-conserving activity. Those with neutral 8-amino acids are called *oxytocins* and have weak ADH activity.

Most mammals have *arginine vasopressin* (AVP), while others such as wild pig and hippopotamus have only *lysine vasopressin* (LVP), and some Suiformes have both. In birds, amphibia, reptiles, and most fishes the active polypeptides are oxytocin, and arginine vasopressin. Other polypeptides are *arginine vasotocin,* AVT *(e.g.,* in lampreys, amphibians, teleosts, reptiles, and birds), *mesotocin* or 8-isoleucine oxytocin (in *Protopterus,* amphibians and birds), *isotocin* or 4-serine, 8-isoleucine oxytocin (in *Polypterus,* holosteans and teleosts), and *glumitocin* or 4-serine, 8-glutamine oxytocin (in elasmobranch, *Raja).*

However, the mode of action of vasopressins is different in the various vertebrate classes. In birds and mammals, the hormone increases the permeability of the renal tubules to water and thus permits more water reabsorption and reduced urine production. In amphibians and reptiles, AVT is also anti-diuretic, but in addition to a small tubular effect the hormone reduces the glomerular filtration rate, probably by constriction of the glomerular arterioles. In anurans, an additional action is to increase permeability of the skin to Na^+ and water ; this, in combination with a sodium pump, favours salt uptake. Another function of AVT is to increase water reabsorption from the bladder. In freshwater fishes (goldfish, eel), AVT increases urine flow and glomerular filtration rate, (Jones *et al,* 1968). Thus, the action on kidneys of freshwater fish is opposite to that in tetrapods, namely to elicit diuresis. Sodium uptake by the gills is increased. In the lungfish *Protopterus,* AVT increases glomerular filtration rate as in freshwater fish, but in addition it enhances Na^+ excretion, *i.e.,* it is natriuretic *(Sawyer,* 1970). Possibly the hormone is important as a diuretic and natriuretic when the lungfish emerges from aestivation. In lamprey, AVT also promotes Na^+ excretion.

Prolactin: In bonyfish, prolactin from the anterior pituitary enhances production of a dilute urine. This hormone is critical for euryhaline fish in migration between fresh and salt water. A hypophysectomised euryhaline fish *Fundulus heteroclitus*, fails to survive well when transferred from sea-to freshwater; but survival is furthered by injection of prolactin. Hypophysectomised freshwater *Fundulus, Gasterosteus* and *Xiphophorus* lose Na^+ rapidly mostly by the gills, but injection of prolactin reduces the rate of Na^+ loss to one-fourth and the fish survive. According to *Lam* and *Leatherland* (1970), prolactin enhances freshwater osmoregulation by both gills and kidneys.

Adrenal Cortical Hormones: Several hormones of the adrenal cortex also function in maintaining salt balance. Aldosterone increases Na^+ reabsorption by the bladder of a toad *Bufo marinus (Sharp* and *Leaf*, 1966). In mammals, aldosterone favours Na^+ retention and K^+ excretion ; Na^+ reabsorption in the distal tubules is enhanced. In a Na^+-depleted dog, low plasma Na^+ increases secretion of aldosterone and corticosterone. Corticosterone enhances K^+ excretion and decreases Na^+ excretion, in salivary glands as well as in kidney. Aldosterone appears to act on toad bladder by inducing the synthesis of some mediator or transport protein, since its effect is blocked by inhibitors of protein synthesis. The action of vasopressin on Na^+ transport by bladder is not so blocked. In birds, glucocorticoids from the adrenal cortex stimulate salt output by the salt glands.

Renin: In mammals, elevated blood Na^+ causes the release from juxtaglomerular cells of the kidney a hormone *renin,* which enzymatically converts angiotensin precursor (angiotensinogen) in the blood to angiotensin I ; this is converted to angiotensin II, which then acts on the adrenal cortex to increase the secretion of *aldosterone.* The latter stimulates reabsorption of Na^+ in the kidney, which in turn favours water retention. Renin seems to be absent from lampreys and elasmobranchs ; it is found more in freshwater fishes than in marine species. When an eel is transferred from fresh to sea water, the renin level in the kidney decreases over a three-week period.

Thyrocalcitonin: Another hormone active in ion regulation is thyrocalcitonin, which is released from the thyroid in the presence of high serum Ca^{2+}. It lowers plasma Ca^{2+} by inhibiting bone resorption. In fish, cells of the ultimobranchial bodies release a polypeptide which lowers blood-Ca^{2+} level of eel and catfish, but not of *Fundulus*.

Neuroendocrine Regulation: Neuroendocrine regulation of salt and water balance has also been observed in crustaceans and insects. In crayfish, ligation of the eyestalks increases water influx and urine production when in fresh water, and increases blood osmoconcentration when in 50 per cent sea water; a brain extract counteracts these effects (*Kamemoto* and *Ono*, 1969). A brief review of what has been discussed above is given again:

(a) Marine invertebrates are in osmotic equilibrium with sea water, but can regulate their ionic composition.

(b) Marine elasmobranches have a low salt content, but are in osmotic equilibrium with sea water because they are capable of retaining urea in their blood.

(c) Marine teleosts are hypotonic and make up for the water loss by drinking sea water and excreting the excess salts by the gills.

(d) Freshwater invertebrates are hypertonic to the water and have mechanisms for actively absorbing salt from the very low concentration in fresh water.

(e) Freshwater fishes and amphibians are in a similar situation and also absorb salt from the water.

(f) Reptiles, birds and mammals are all terrestrial and have impermeable integuments and some other devices for the conservation of water.

(g) Some hormones control osmotic and ionic balance in both invertebrates arid vertebrates.

5

Significance of Homeostasis

The term 'homeostasis' consists of two parts: *homeo* which means "like" or "similar", and stasis meaning "a standing" or "a state of equilibrium." Thus, *homeostasis* is defined as the tendency to maintain stability or equilibrium within an organism or a social group. The term homeostasis is used by physiologists to mean maintenance of constant or static conditions in the internal environment.

However, we should differentiate between homeostasis and *cybernetics*. The term 'cybernetics' was coined by *Norbert Wiener* (1948) and is defined as "the science of controls." Homeostasis includes the mechanisms responsible for the maintenance of the internal environment in the living organisms and also the tendency to maintain equilibrium between the internal and the external environment of the organism or social groups.

It is also necessary to differentiate between homeostasis and haemostasis. The term *haemostasis* means the "control of bleeding" and depends in part on appropriate vascular responses at the site of injury.

Brief History: The French Physiologist *Claude Bernard* (1813-1878) was the first to observe the importance of homeostasis in animal function. As a physiologist he was concerned with the *internal environment* ("milieu interieur") of the living organism and described that internal environment resisted change and seemed to remain fixed ("La fixite").

According to him, "all the vital mechanisms, however varied they may be, have only one object, that of preserving constant conditions of the life in the inner environment". In other words, *"constancy of the internal environment"*, as *Bernard* (1859) phrased it, has been found to be a universal phenomenon in living animals and plants, allowing them to survive in different habitats including stressful and varying external environments.

In 1929, the American physiologist *Walter B. Cannon* published an article entitled "Organisation for Physiological Homeostasis". He suddenly was struck by the remarkable constancy of the internal environment. But in the French edition of his book, *The Wisdom of the Body, Cannon* made it clear that *Bernard* deserves priority for emphasising the role of the inner environment in the establishment and maintenance of steady states in the body. In other words, it was the Bernard's original idea of the constancy of the internal environment that *Cannon* adopted, developed, and then neatly packaged under the label of *homeostasis.*

The following lines of *Cannon* (1932) may be quoted. "The coordinated physiological processes which maintain most of the steady states in the organism are so complex and so peculiar to living beings— involving, as they may, the brain and nerves, the heart, lungs, kidneys and spleen, all working cooperatively—that I have suggested a special designation for these states, *homeostasis.* The word does not imply something set and immobile, a stagnation. It means a condition—a condition which may vary, but which is relatively constant."

Langley (1965) in his brilliant monograph "Homeostasis"

writes: "*Bernard* and *Cannon* and many other physiologists were impressed by the self-regulating negative feedback systems which serve to maintain the constancy of the internal environment of living organisms. This is termed homeostasis. But homeostasis is not limited to animals, it is applicable to all living things. Engineers have long used such systems. The principle may even be applicable in sociology, economics, and ecology".

Homeostatic Mechanical Theory

The regulatory processes are based primarily on the principle of feedback or *feedback mechanism* or *servomechanism* (as the engineers call it). The essential components of such a mechanism include: (1) a sensor, (2) a circuit, (3) an effector, and (4) a shut off device. In servomechanism, the terms *input* and *output* are used. If we take an example of thermostatic regulation of room temperature, the working of the basic elements of homeostasis may be explained as follows:

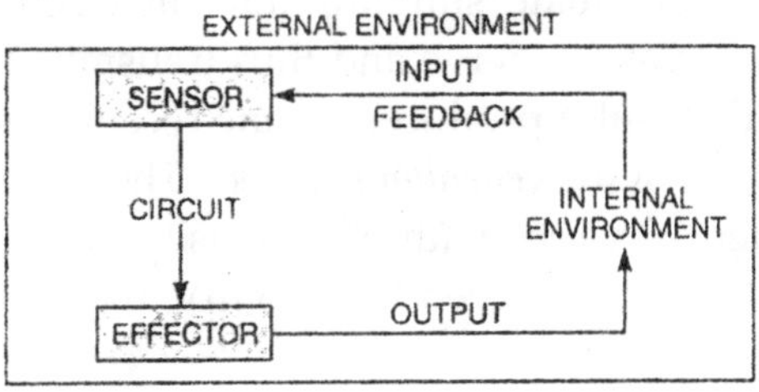

Basic elements of homeostatic mechanism.

If we want to keep the house temperature between 22°C and 25°C, the *sensor* (thermostat) must respond when the temperature of the house rises above 25°C, or falls below 22°C. In the first case when it rises above 25°C the sensor responds by closing a circuit which activates the air conditioner. As a result, air at a temperature considerably below 22°C is blown into the house. The internal temperature of the house will, accordingly, fall. Similarly, if the temperature of the house falls below 22°C, the thermostat responds by closing a circuit which activates the furnace to produce hot air. In this example, the temperature of the room is the *input,* the heat produced by the furnace (or the cold air generated by the air conditioner) is the *output.*

Each living organism maintains its internal environment in a similar manner. There are *sensors* called *receptors* ; there are *circuits* which in some instances are composed of *nerves,* but in other cases utilise different mechanisms, for example, the circulation of hormones in the blood, and there are *effectors* which can alter the internal environment and return it to the normal narrow range despite wide variations in the external environment. In the living organism receptors are *biological transducers (Langley,* 1965). A receptor is specialised nervous tissue sensitive to a specific change in its environment.

Receptors generally respond only to a specific type of stimulus. For example, a receptor may respond to changes in temperature, but not to changes in light or pressure. As biological transducers, receptors are capable of converting the energy of the adequate stimulus into an electrical impulse which then activates a nerve and thus transmits the message. The electrical impulse produced by the receptor is termed the *receptor potential* or the *generator potential.* The magnitude of the receptor potential varies with the intensity of the stimulus. If the potential does not exceed 10 to 15 mV, the nerve is usually not activated.

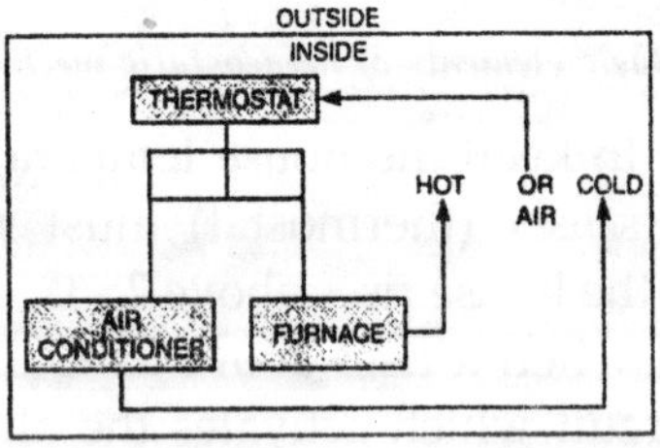

Control of house air temperature by simple feedback mechanism.

In homeostatic mechanisms we deal with messages. The first message which indicates a difference between the external and internal environment influences a transducer. The latter converts the energy of that message to a convenient form. In that form a second message is transmitted to a device which can do something to oppose the influence of the external environment (*Langley*, 1965).

Characteristics of Homeostatic Mechanisms: All the homeostatic control mechanisms have certain characteristics in common, which are as follows:

1. *Negative Feedback:* Most control systems of the body act by negative feedback. For example, in the regulation of CO_2 concentration, a high concentration of CO_2 in the extra-cellular fluid increases pulmonary ventilation. This in turn decreases CO_2 concentration because the lungs then excrete greater amounts of CO_2 out of the body. In other words, the high concentration causes a decreased concentration, which is negative to the initiating stimulus.

Therefore, in general, when some factor becomes too high or too low, a control system initiates *negative feedback,* which consists of a series of changes that return the factor towards a certain mean value, thus maintaining homeostasis.

2. *Gain of a Control System:* The degree of effectiveness with which a control system maintains homeostasis is determined by the *gain* of the negative feedback. The gain of the control system is calculated by the following formula:

Gain = Correction divided by error.

For example, a large volume of blood is transfused into a person whose barorecepter pressure control system is not functioning, and the arterial pressure rises from the normal level of 100 mm Hg to 175 mm Hg. Then, if the same volume of blood is injected into the same person when the baroreceptor system is functioning and this time the pressure rises only 25 mm Hg. Thus, the feedback control system has caused a "correction" of -50 Hg, that is from 175 mm Hg to 125 mm Hg.

There remains an increase in pressure of +25 mm Hg, called the "error", which means that the control system is not 100 per cent effective in preventing change.

The gain of the baroreceptor system for control of arterial pressure is -50 divided by +25, or -2. That is, an extraneous factor that decreases or increases the arterial pressure does so only one third as much as would occur if this control system were not present. But the gains of some other control systems are much higher than that of the barorecepter system. For example, the gain of the system controlling body temperature is about -33, which indicates that the temperature control system is much more effective than the barorecepter system.

3. Adaptive control system: In some cases more complex types of control systems are used. For instance, some movements of the body occur so rapidly that there is not enough time for nerve signals to travel from the peripheral parts of the body to the brain and then back to the periphery again in time to control the movements. Therefore, the brain uses a principle called *feed-forward control* to cause the required muscle contractions. Then, sensory nerve signals from the moving parts apprise the brain in retrospect whether or not the appropriate movement as envisaged by the brain actually has been performed correctly. If not, the brain corrects the feed-forward signals that it sends to the muscles the next time the movement is required. Then, once again, if still further correction needs to be made, this, too, will be done for subsequent movements. This is called *adaptive control,* which, in a sense, is delayed negative feedback.

Examples of Homeostatic Mechanisms

Numerous examples of homeostatic mechanisms may be given right from subcellular and cellular level to organism, population and ecosystem level. The following examples from human physiology are important.

1. *Regulation of Body Fluids:* The volume and composition of body fluids are homeostatically maintained with almost complete precision. The body fluids consist of *intracellular fluid* (fluid within the cells) and *extracellular fluid* (blood and interstitial fluid). As blood passes through the capillaries, continuous exchange of extracellular fluid also occurs between the plasma portion of the blood and the interstitial fluid that fills the intercellular spaces, the spaces between the cells.

Blood contains a wide variety of substances—glucose, potassium, sodium, calcium, *etc.*, each of which is regulated by a feedback mechanism. Although a number of factors work for the regulation of the volume of the body fluids, two major factors are: *(a)* the control of urine formation, and *(b)* the control of fluid intake.

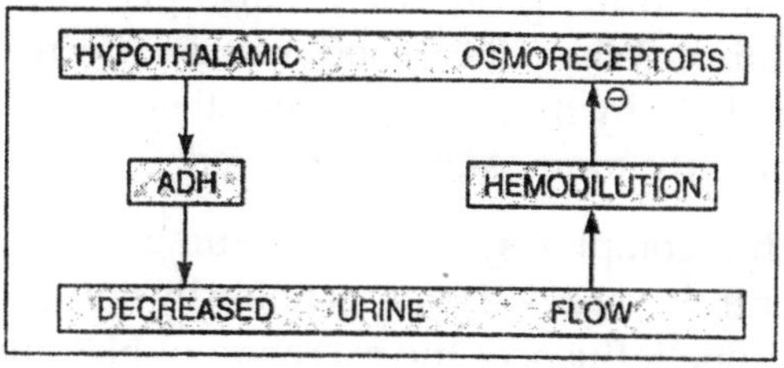

Control of urine volume.

(a) *Control of Urine Output:* There are two types of sensors that are necessary to fluid volume control—*osmoreceptors* (that evaluate the concentration) and *volume receptors* (that evaluate the change in volume). The former are located in the hypothalamus. When the organism is dehydrated, the osmotic concentration of the blood increases and therefore hypothalamic osmoreceptors respond. As a result, messages are sent to the posterior lobe of the hypophysis and a hormone called the *anti-diuretic hormone* (ADH) is secreted. It is carried by the blood to the kidneys where it acts to diminish the formation of urine. Thus urine volume is reduced, and if fluid intake remains normal, fluid will accumulate in the body thereby returning the osmotic state of the blood to normal. Thus, there is a simple but quite effective negative feedback mechanism, which regulates the volume of urine vis-a-vis osmotic state of the blood.

It should be noted that in this feedback mechanism the sensor is sensitive only to the concentration (osmotic pressure) of the blood. But there is also a feedback mechanism that responds to the volume of the blood. For example, following haemorrhage there is a sharp decrease in urine formation. Such volume receptors have been located in the walls of the atria and great veins. As they are stretch receptors, they respond when the volume of blood increases. These messages are inhibitory and cause a reduction in the release of ADH. Pressoreceptors in the great arteries also play a role in body fluid regulation. When the blood volume decreases, fewer impulses are fired by the *pressoreceptors,* there is less inhibition and consequently more ADH is secreted.

Through a complex system the volume of the blood also exercises control over *aldosterone* secretion by the adrenal cortex. The aldosterone influences the excretion of Na^+ and water by the kidneys. The sensors are thought to be the specific cells in the blood vessels of the kidneys. These cells themselves secrete a substance called *renin.* In this case, then, the receptors are also effectors. That is, instead of sending message to other cells that respond, they respond themselves. The renin secreted into the blood reacts with a constituent of the blood called *angiotensinogen* to form *angiotensin.* Angiotensin then is carried by the blood to the adrenal cortex where it stimulates the secretion of *aldosterone.*

The aldosterone is carried by the blood back to the kidney where it decreases the excretion of Na^+ and water. The more the kidney cells are stretched, the less renin they secrete. As the blood volume decreases, the volume receptors secrete more renin, more angiotensin is formed, the adrenal cortex secretes more aldosterone, the kidneys retain more Na^+ and water, and therefore the blood volume increases, thus completing the cycle.

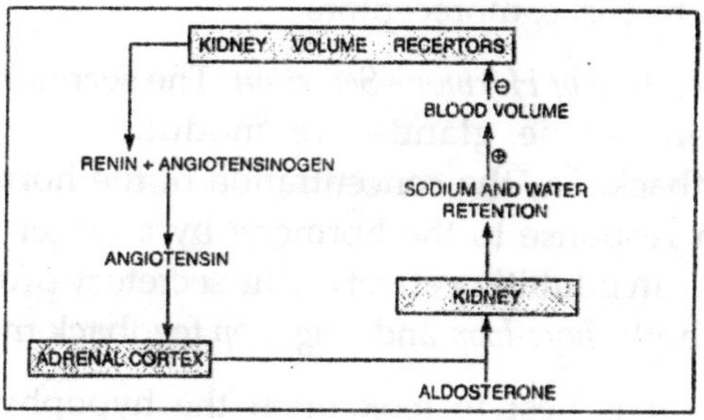

Control of body fluid volume.

Thus with at least three mechanisms controlling the secretion of ADH (1.) the osmoreceptors in the hypothalamus, (2.) the stretch receptors in the great vein and atria, and (3.) the pressoreceptors in the arteries) and at least one regulating aldosterone secretion, it is no wonder that the urine output is so precisely attuned to body fluid needs.

(b) *Control of Fluid Intake:* We are familiar with the fact that in hot weather, due to the loss of water through sweating, we feel thirsty and accordingly our fluid intake is increased. The sensation of *thirst* has at least two components: *(i)* a local response represented by drying of the mucous membrane of the mouth and throat, and *(ii)* the hypothalamic drinking centre. The cells that make up this part of the hypothalamus an the sensors of the controlling feedback mechanism of thirst

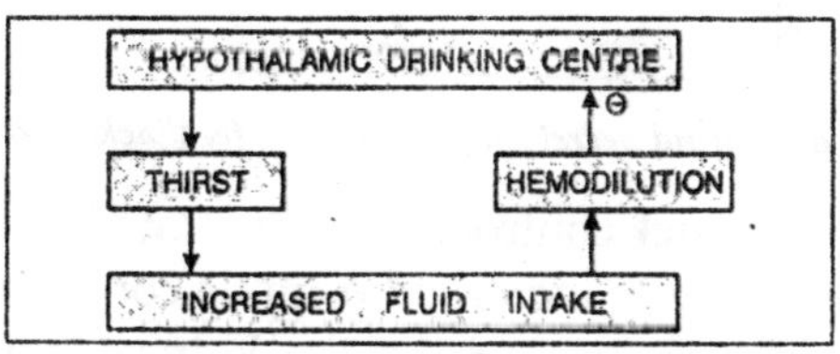

Control of fluid intake.

They are osmoreceptors and monitor the osmotic concentration of the blood. As a result of thirst, fluid intake is increased; this dilutes the blood thereby decreasing the stimulation to the osmoreceptors.

2. *Regulation of Hormone Secretion:* The secretory activities of endocrine glands are modulated by negative feedback, *i.e.*, the concentration of the hormone itself, or a response to the hormone by a target tissue, will have an inhibiting effect on the secretory process. There are both *short-loop* and *long-loop* feedback mechanisms.

Perhaps it is best to look upon the hypophysis-thyroid interrelationship as the fundamental homeostatic mechanism and the hypothalamus as a part that can alter the setting of that fundamental mechanism (*Langley*, 1965).

The anterior lobe of the hypophysis secretes the *thyroid stimulating hormone* (TSH), which stimulates the thyroid gland to secrete *thyroxin*. But thyroxine is inhibitory to the anterior lobe of hypophysis. As a result, less TSH is secreted. Through this short loop negative feedback mechanism, the thyroid secretion is controlled. Similarly, the level of blood calcium regulates the activity of the parathyroid gland.

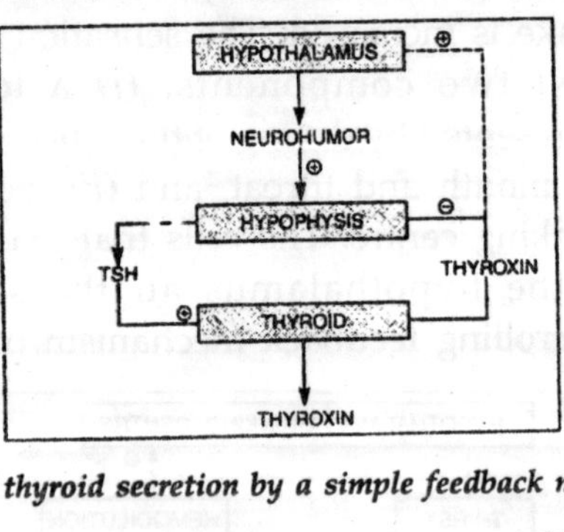

Control of thyroid secretion by a simple feedback mechanism.

Negative feedback control is also illustrated by the secretion of ADH in response to increased osmotic concentration of the blood. The ADH upon reaching the kidney increases water reabsorption thus counteracting the loss of fluid. As a result, plasma osmolarity is stabilised, and, in turn, so is the secretory activity of the ADH-producing tissue of the posterior lobe of the hypophysis.

The negative feedback signal acting to slow the output of an endocrine tissue may be a physiological response to the hormone secreted by that tissue *(e.g.,* reduced osmolarity in the ADH loop ; reduced blood glucose levels in the insulin loop) or it may be a rise in concentration of a second hormone, the secretion of which is stimulated by the first hormone *(e.g.,* high *estrogen* levels inhibit production of the *gonadotropic hormone* that stimulates estrogen production). Thus the negative feedback signal may be simply the reduction of a stimulus to the endocrine tissue, or it may be an active inhibition of secretion of the hormone.

3. Other Examples: Many other examples of homeostatic mechanisms may be found in physiological processes such as (i) control of blood pressure, (ii) control of respiration, (iii) control of blood glucose level, and (iv) control of body temperature, etc.

4. *Homeostasis and Behaviour:* The role played by *behaviour* in the maintenance of homeostasis varies considerably with species and with circumstances. For example, *drinking behaviour* is essential for the maintenance of homeostasis in many organisms, the physiological mechanisms of water conservation being unable to prevent lethal dehydration after prolonged water deprivation. However, some species such as the budgerigar *(Melopsittacus undulatus)* and the mongolian gerbil *(Meriones unguiculatus)* are able to survive indefinitely without water. Because of the great efficiency of their water conservation mechanisms they can live on the water contained in the seeds they eat *(McFarland,* 1981). Others, such as aquatic species need no special behaviour to obtain the water necessary for the maintenance of homeostasis.

In some aspects of homeostasis the role of behaviour is negligible. Experiments involving surgical manipulation with the physiological mechanisms responsible for homeostasis show that animals often possess the capacity for appropriate behaviour, even though it is not manifested in their normal lives. For example, when the thermal homeostasis of rats is interfered with by removal of the thyroid gland, they respond if given the opportunity by the experimenter, by building warmer nests and by showing other forms of behavioural thermoregulation *(Curt Richter,* 1943).

In a similar way, removal of the adrenal gland, which is involved in salt balance, causes rats to shift their food preferences towards more salty food and water. In other cases there may be no special organs (such as thyroid and adrenal glands mentioned above) involved. Rats fed on a vitamin-deficient diet are able to select food containing the required vitamins, even though they are unable to taste the presence of vitamins in the food. It seems that they are able to learn which food made them feel better *(Rozin* and *Kalat,* 1971). But how does the rat know which food to select at a particular time? The answer is that for some constituents of diet such as water, sodium, and possibly glucose, there is different monitoring by taste and by sensory processes in the brain that detect the presence of each of these substances in the blood. For other dietary elements such as the vitamins and minerals, there is no direct monitoring and the animal learns what to eat in order to avoid sickness.

It may be mentioned that *homeostatic requirements* place behavioural demands upon animals. The traditional view of *motivation* is built upon a simple feedback principle. A change in the animal's internal state is sensed by the brain and leads to a drive to perform the appropriate behaviour. The classical view of hunger and thirst as homeostatic drives implies that feeding and drinking are initiated as a result of monitored changes in the animal's physiological state. Feeding and drinking are said to be under negative feedback control because food and water intake serve to diminish the deviations in physiological state that initiated the behaviour.

However, a distinction has been made between homeostatic aspects of behaviour, such as feeding, drinking, and temperature regulation, and non-homeostatic aspects of behaviour, such as aggression and sexual behaviour. According to *McFarland* and *Nunez* (1978) and *Toates* (1980), the control of sexual behaviour is not fundamentally different than that of feeding or drinking. The interactions among the physiological processes that give rise to behaviour are complex.

Physiological processes may become attuned to the normal routine. In humans, for example, the liver may cease to mobilise glycogen just before a meal is due. This leads to a fall in blood sugar level in anticipation of the increase that will occur when the meal is digested. Experiments have shown that such physiological adjustments can undergo conditioning in relation to the time of day.

It may be concluded that the concept of homeostasis has had great influence upon the development of thought in animal behaviour research. This is partly because an animal's physiological requirements and interactions among physiological processes are bound to have a great effect upon its *motivation* and *motivational state,* and partly because homeostasis is an excellent example of the *feedback principle,* by which the behaviour is partly controlled by its own consequences.

6

The Tissues

Histology is the study of tissue sectioned as a thin slice, using a microscope. It can be described as microscopic anatomy. Histology is an essential tool of biology. Histopathology, the microscopic study of diseased tissue, is an important tool of anatomical pathology since accurate diagnosis of tumours and other diseases usually requires histopathological examination of samples. The trained technicians who perform the work of processing, thin sectioning, and staining these tissues are Histologic Technicians (HT) and Histologic Technologists (HTL). Their field of study is called histotechnology.

The study of Histology begins with the cell and tissues. Each of the four primary tissues of the body has characteristics structural and functional features. Formerly, the study of the basic tissues used to be called 'histolgy' (histos = tissue) and the structure of organs and systems 'microscopic anatomy'.

Plant Tissues

Tissues are combinations of cells, organs are functional unities of an organism. The basic organs of the phanerogamous plant are leaf, stem and root. Organs consist of different tissues, the leaves for example of dermal tissue, assimilation tissue and vascular tissues. A tissue again may have differently structured cell types. The vascular tissue contains thus the cells of the xylem and those of the phloem. Meristems are characterised by cell divisions, while this is an exceptional feature in permanent tissues. The latter are as a rule differentiated and often specialised. Depending on their functional properties can they be grouped into the following categories:

- Dermal tissues (epidermis, cork, bark)
- Ground tissues (parenchyma)
- Assimilation tissues (palisade parenchyma, spongy mesophyll)
- Collenchyma and sclerenchyma tissues (collenchyma, sclereids, fibres)
- Vascular tissues (vessel elements, xylem, phloem)

Meristems : Cormophytes produce new cells throughout their life and form new organs at regular intervals (some tallophytes do the same). All cells of an embryo display the same activity of division, but as soon as a certain state of differentiation and thus a certain size has been reached, the production of new cells is restricted to special tissues, the meristems.

Typical meristems are found at the tips of all stems and roots. They are called apical meristems. Those of the shoot are usually protected by involucral leaves and the whole complex forms a bud. The apical meristems are the cause of tip growth that is one of the most striking features of vegetative growth. This contrasts with animal growth, that is an allometric process, i.e. all parts of the body grow in a weay that maintains the body's overall proportions. As soon as an animal has reached its final size, growth ceases. This, however, does not mean that no more cell divisions are taking place. Cell divisions occur all the time, especially in epithelia and in blood stem cells. But this division activity has always the function of replacing old or damaged cells. No such repair occurs in plants.

If a leaf or flower or any other part is damaged, it is neither replaced nor repaired. Instead, new organs grow, and this again does not happen in animals. The cells of a meristem have a potentially unlimited ability to divide. But division is controlled. Many plants, for example, follow the principle of apical dominance, where the activity of the lateral meristems is suppressed while the tip is growing. In this case, information is exchanged between the two tissues in question. If the tip of the main axis is cut off, suppression is abolished and the lateral buds start to sprout.

Many plants like gymnosperms and dicots have extensive lateral meristems (the cambium, the vascular cambium, and the cork cambium) that give rise to growth in girth, also called secondary growth. Their activity may fluctuate in the course of the year and is reflected by annual rings. Other plants, like monocots, have intercalary meristems. They are actively growing meristems that differ clearly from apical meristems and are located between more or less differentiated tissues usually at the base of each internode.

The formation of secondary meristems shows that differentiated cells can reverse their state and go back to a meristematic existence. Their ability to divide is thus not lost, though it is not used unless an adjustment to changed circumstances requires it. The ability of indeterminate growth is widely exploited by the use of cuttings in horticulture.

A meristem does not only include the initial cells and their immediate progenitors, but also some parts of the shoot. The situation is similar in root meristems, although their anatomy is different. The shoot apex is protected by the involucral leaves. In roots, this function is exerted by the root cap. Its cells are produced by a specialised root cap meristem and are progressively replaced towards the tip. These cells have to have a high turn-over, since the outmost cell layer is easily damaged by soil particles as a result of the growth movement of the root.

Procambium and lateral Meristems : It is normally

distinguished between primary and secondary growth of the shoot. The first is the phase of plant development that gives rise to new organs and to the basic plant form, the latter brings about growth in girth and the formation of new vascular tissues. The procambium is the meristematic tissue that produces the primary vascular tissues: xylem and phloem. It develops directly beneath the growing tip next to new leaf primordia. The development of new leaves is thus tightly connected with that of new vascular tissues.

The vascular tissues of leaves are also called leaf veins. The cells of the procambium are generally combined in cords. They form an extension of the vascular tissues into the growing tip and do thus provide the connection of the newly developed organ with the conductive systems of the plant. The cells of the procambium are elongated and become even more so in the course of their development. The volume of their vacuoles increases considerably, lending them a lighter and more transparent appearance than their neighbouring cells. This is decisive for their differentiation into xylem or phloem cells.

Epidermal tissues : The epidermis has numerous functions:

- Protection against various chemical and physical influences, against being fed upon by animals and against infestation by parasites
- Protection of the plant against desiccation
- Participation in gas exchange, in secretion of metabolic compounds and in absorption of water
- Site of receptors for light and mechanical stimuli that help to transform signals from the surrounding to the plant

The epidermis has accordingly a number of differentiated cell types to serve the various functions. Variations typical for certain species and different organisations of the epidermis in the miscellaneous plant organs add to the number of different cells. Three main types exist:

1. The basic epidermis cell
2. The cells of the stoma complexes and
3. The trichomes (gr.: trichoma = hair), epidermal attachments of varying shape, structure and function

The basic epidermis cells, i.e. the least specialised cells constitute the largest group of dermal cells. They seem either polygonal or elongated in top view. Their walls are often wavy or sinuate. It is unknown, what induces this shape during development, since the explanations given by the existing hypotheses seem insufficient. Elongated epidermis cells can be found at organs or parts of organs that are elongated themselves, like stems, leaf petioles, leaf veins or leaves of monocots. The epidermises of the leafs upper- and undersurface may have different structures. The shape of the cells, the thickness of the walls as well as the distribution and number of specialised cells (guard cells and trichomes) per area may all vary.

Wide varieties of different cell shapes may even exist in species of a single family, e. g. in the Crassulaceae family. The wall of epidermal cells that constitute the leaf's surface is often thicker than the other walls. This can be particularly well observed with the epidermis of conifer needles and that of xerophytes (plants living in dry habitats). Aquatic plants have usually thin walls. The wall of many seeds becomes stronger during ripening and may fill nearly all of the cell's lumen so that the protoplast is driven out and degenerates. The basic epidermal cells of most species contain no chloroplasts. Some ferns and several aquatic or shade plants are exceptions.

The epidermis is more often than not built from a single cell layer, though multi-layered, water-storing epidermises that evolved from initially single-layered tissues by periclinal division have been shown among the species of several families (Moraceae: most Ficus-species, Piperaceae: Peperonia, Begoniaceae, Malvaceae and others). Epidermis cells secrete a cuticle, that covers all epidermal surfaces like an uninterrupted film. It may either be smooth or structured by bulges, rods, filaments, folds, or furrows.

The folding of the cuticle is nevertheless not always caused by cuticular rods or filaments. There are cases where it is merely the expression of the bulging of the cell walls below. It may be necessary to analyse cross-sections of the cuticle and the underlying epidermal cells to find the actual cause. In some cases, like that of the tomato fruit, the cuticle is pigmented with carotinoids. Often, additional waxes, oils, resins, salt crystals and (hydrophilic) mucilage are excreted. The latter is especially common in developing seeds. The excretion of wax inhibits the moistening of the leaves more than the cuticle itself. These waxes do often have an intrinsic structuring. A thick layer of wax lends a whitish appearance to the plant surface. It has two effects: diminution of water-loss and enhancement of reflection of the sunlight, thus protecting the plant from too much heat.

Stomata and guard cells : The stomata accomplish an important function of the epidermis. Their complete functional unit is the stoma complex. It is composed of two chloroplast-containing guard cells that enclose a pore (the actual stoma), as well as of two to four subsidiary (or accessory) cells devoid of chloroplasts. Cross-sections show that the guard cells have walls of different thicknesses. They sit at the top of an intercellular space, that is in communication with other intercellular spaces of the respective tissue. The stoma can be opened or closed according to demand. In this way, the guard cells regulate both transpiration and CO_2- uptake.

Water-content and CO_2- concentration within the plant tissue act as regulators for the state of opening. The guard cells control the size of the stoma by changing its shape: more about the mechanism. Stomata occur at all plant surfaces above ground. Their number approaches 100-300 per mm^2, but there exists a wide variation in number and distribution. They are lacking in roots, in the epidermis of the chlorophyll-free shoot of the parasitic species Monotropa hypopitys and Neottia nidus-avis as well as in some submerse living aquatic plants, while they are normal in others. They do, too, happen to be present in some coloured or white petals, though they have often lost their function here.

The paralleled veined leaves of most monocots, some dicots and the needles of conifers have stomata that are grouped in parallel rows. Stomata are produced by the guard cell mother cells, that again are laid out within the epidermis in regular intervals. The initial of the stoma complex is the smaller of the two progenitor cells. It contains more plasma. Both guard cells stem from it and are formed by unequal division, while the production of the subsidiary or accessory cells may occur in many different ways. The development of a leaf's stoma happens mostly asynchronously. It follows the successive differentiation of the single tissues in leaves with parallel venation. Differentiation proceeds in a basipetal manner, i.e. it starts at the tip of the leaf and continues towards its base.

Netted leaves have stomata of different developmental stages spread mosaic-like across the whole lamina. Morphological variations of stoma complexes can be expected, since different ways of their development exist. The guard cells of gramineous plants are a typical exception. They are often shaped like dumb-bells. The hydathodes, that can often be found at the end of vascular bundles are other derivatives of stoma complexes. Their guard cells do still look like those of normal stoma complexes, but they cannot be closed any more.

Water secretion by the hydathodes is called guttation. Characteristic hydathodes occur especially at the leaf margins of Garden Nasturtium (Tropaeolum majus), Lady's Mantle and at the leaves tips of many grasses. Salt, sugar and organic compounds that are dissolved in the guttation water chrysalis after evaporation of the water at the site of outflow. Typical examples are the lime secretions of Saxifrage-species (Saxifraga) and the salt glands of halophytes.

Trichomes : Epidermal attachments of various shape, structure and function are called trichomes. They protect and support the leaf, produce glands in the form of scales, different papills and, in roots, often absorbing hairs. They stem exclusively from epidermal cells. Often a trichome is formed by just one cell though sometimes several cells are involved. Trichomes have to be distinguished from multi-cellular trichomes (like spines) and from short shoots (like thorns), since these do contain not only cells of epidermal origin but of other tissues, too.

Hairs can be found at plant surfaces in manifold shapes. They can be single- or multi-celled, branched or unbranched, living or dead. Their walls may be strengthened by silicate, calcium carbonate or other encrustations giving the respective hairs a bristle-like appearance. Such stiff bristles (at Boraginaceae and Cruciferaceae for example) protect plants from being eaten away by animals. Many hairs, particularly the strongly branched ones, help avoiding transpiration losses. It is well known that plants from dry habitats are either succulent (and thus have a thick cuticle) or densely covered with silvery hair. 120 hairs per mm^2 were counted at Stachys lanata (woundwort). Their microscopic analysis shows strongly branched and dead hair cells. These hairs offer three advantages:

1. The lumen of dead cells is air-filled. It gives them a silvery-whitish appearance. A large proportion of the incoming light is thus reflected. A thick layer of wax has the same effect.
2. Only weak circulation takes place at the leaf surface, thus reducing water loss to a minimum.
3. Dying of the hair cells reduces the surface area, where water could be lost drastically. If the cells were living, too a high loss of water loss would occur due to the largely increased surface of the branched hair cells.

Endodermis : The epidermis develops from the outer layer of the apical meristem. The origin of the rhizodermis is much less evident. Depending on the species, it may during evolution either have derived from the root cap or the primary bark. The root hairs that absorb water and nutriments are an important feature of the rhizodermis. With some species of the Leguminosae family, they participate in addition in the recognition and uptake of nitrogen-fixing bacteria (belonging to the genera Rhizobium). They originate in a zone with high cell division activity. A root hair develops usually from a small extrusion (a papilla) at the apical end of a cell.

In many plant species, all cell types of the rhizodermis are able to develop root hairs, while some species have specialised cells (trichoblasts) for this purpose. They are arranged at a regular pattern on the surface of the rhizodermis. The origin and the growth (elongation) of hairs in the different root parts occurs synchronised. The oldest hairs are the longest, while those nearest the root tip are shortest.

Root hairs can become as long as 80-150 um and as wide as 5-17 um in girth. They are highly vacuolate when fully differentiated. Only rarely do they consist of many cells (adventive roots of Kalanchoe). Root hairs are shortlived (a few days) and are continually replaced by new ones. The formation of root hairs helps to increase the size of the absorbing root surface drastically. Root hairs are always in close contact with soil particles. They are often grown together and it is thus impossible to isolate a root that has been grown in the soil without impairing it. As a rule, root hairs produce carbohydrate-containing mucigel. It generates a rhizosphere, where nutriments can be processed before uptake by the plant.

Another specialised cell in the root is the endodermal cell. The endodermis forms a cylinder around the vascular cylinder in the roots. The endodermal cells have a Suberin layer called the casparian strip embedded into their cell walls. This Suberin can be stained using a special dye and appears when viewed under fluorescent light. The casparian strip appears as light spots in the endodermis in a longitudinal section through the root, but the Suberin is laid down in a continuous layer that cannot be seen in longitudinal sections. The Velamen radicum is a special epidermis occurring only in aerial roots. It is an epidermis of several cell layers that is produced by the aerial roots of Araceae and tree-dwelling orchids.

Most of its cells are dead and store water like a sponge. The Velamen looks silvery-white when dry, since the cells are filled with air. In a moist state, the chloroplasts of the underlying tissue shimmer through and the roots look green.

Secondary Dermal Tissues : Secondary dermal tissues are usually found in older shoots and roots, because these are the organs where secondary growth takes place. Only rarely and only in their initial stages do secondary tissues occur in bryophytes, ferns and monocots.

They stem from the phellogen, a secondarily developed cambium. The cells of the phellem are called cork cells, they are generated centrifugally, are non-living and have suberised cell walls. The phelloderm consists of cells given off towards the inside of the phellogen, forming the inner part of the periderm. Periderm is the collective term for all three types of tissue.

Cork layers are neither permeable by water nor gas and do only rarely constitute homogeneous structures. The cracked surface structure of the bark that typical for certain species is a visible expression of this. The phellem is often interspersed with groups of parenchyma cells of elongated appearance, called lenticels. These can be observed on the surface of elder (Sambucus niger), a typical specimen of basic botanical courses.

The lenticels serve to permit gas-exchange between the metabolically active cells below the bark and the atmosphere. They are not only found in the shoot, but in roots, too. Cork is rarely found within primary tissues, the leaves of Tabernnaemontana pachysipon are an exception. They have little islets of cork at the undersurface of their leaves. The totality of all tissues produced by the phellogen, both primary and secondary ones as well as some other types, is called bark. You could simplify things by saying that all tissue outside the cork cambium is bark.

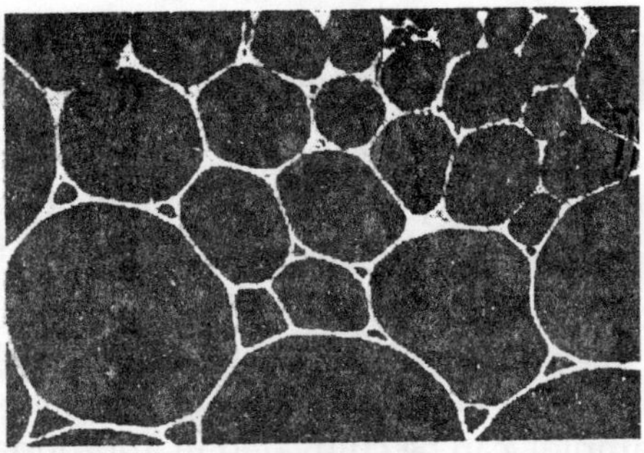

Parenchyma cells and relatively large intercellular spaces in a cross-section through the stem of Geum urbanum.

Parenchyma : Parenchyma or ground tissue cells are always alive, usually isodiametric and only seldom elongated. The pith of the shoots, the storage tissue of the fruits, the seeds, the roots and other underground organs are all parenchyma tissues, as is the mesophyll (the assimilation tissue of leaves). The mesophyll is reviewed separately due to its fundamental function in assimilation and its resulting importance for plant nutrition.

The entire function of the parenchyma seems often to be in the filling of the gaps between more specialised cells. This is expressed by the term ground tissue, which means the mass of cells into which more specialised cells, like vascular tissues or seeds are imbedded. Parenchyma cells are not differentiated, neither morphologically nor physiologically. It is not uncommon that they contain chloroplasts. Sometimes, especially when grown without light, leucoplasts or incompletely structured chloroplasts can be found.

Although hardly any cell division takes place in differentiated parenchyma, the cells keep their ability to divide: they make up the bulk of cells participating in wound healing and regeneration of plant organs. Consequently, the cells could be said to be in a kind of waiting position. They form a pool of cells that can be activated when required, for example during ontogenesis and at unusual incidents. The meristem has already been defined as the tissue, where most cell divisions take place.

The parenchyma could analogously be described as a basic tissue, whereof very different tissues develop during ontogenesis.

The impact of a cell's position has already been mentioned. Whether or not a cell divides depends on its position (information of position) within the tissue. In the case of parenchyma cells, the position has a decisive influence on their further development. Parenchyma tissues (from the pith of the shoot, for example), can be cultivated in appropriate synthetic culture media and are thus prompted to divide. They keep their capacity for cell division for several decades, if parts of the culture are transferred to fresh media at regular intervals. The cells stay in an undifferentiated state. They form aggregates, the calli (sing. callus). The cultures themselves are called tissue- or callus-cultures. The calli remain colourless, when kept in darkness, while they become usually green upon light-exposure. Addition of suitable growth hormones (phytohormones) leads to the development of shoots or roots.

The metabolism of plants produces a vast variety of substances, some of which are stored for subsequent use. In general, parenchyma cells of different organs above or underground serve as depots. The chemical features of these substances will be reviewed later. It is, at present, sufficient to know that - for cells - basic differences between small and large molecules exist. The majority of small molecules is made up by inorganic ions, organic acids and their salts, sugars, a number of nitrogen-containing compounds (like amino acids or alkaloids) and several others.

Small molecules are generally stored as solutes within the vacuole. They are sometimes crystallised and these crystals are either stored within the vacuole or the plasma. Large molecules, or macromolecules, are, for example, starch and proteins. Macromolecules do rarely find a way to the central vacuole. They are stored as microscopic particles within the plasma. Sugar and starch are the most important primary products of photosynthesis. It is commonly known that the parenchyma of fruits may contain large amounts of different sugars. The simultaneous storage of organic acids in the same cells assigns the fruits their specific, often sour-sweet taste. Starch depots can be found in seeds (within the endosperm, a parenchyma tissue), in roots, tubers (dahlia, potato) and others.

Proteins are frequently stored in seeds, usually in cells specialised for this function. A wheat grain, for example, has starch in its central endosperm cells and proteins in the form of aleuron granules at the periphery. These proteins, the storage-proteins, belong to a certain class of proteins. Their chemical composition and structure is dependent on the respective species. When germinating, part of the macromolecules of the endosperm are degraded and used for the organisation of the new plant. After mobilisation of its contents, the endosperm usually dies.

Besides the already mentioned chemical compounds, parenchyma cells store huge amounts of water. This is especially striking in succulent plants, like cactus, Aloe, Agave and Aizoaceae (Mesembryanthemum family). The water content of single parts of a bamboo shoot and its content of parenchyma cells in per cent can be correlated linearly. Certain organs of plants of our regions do also contain more water than others. Among these are the flesh of fruits, onions, buds and all fleshy parts above ground. Most parenchyma tissues, particularly those with storage functions have only small, if any intercellular spaces at all. The thin cell walls put up hardly any resistance against the extension of the cell contents due to the high osmotic pressure (turgor) of the single cell, so that intercellular spaces cannot be formed in the first place.

Aerenchymas : The case is different in a special type of parenchymatic cells characterised by thicker walls. By splaying of such cells or tissues, air filled cavities are generated. In the shoot and the roots, these cavities are usually arranged tangentially or radially. The splaying is caused by tension differences in neighbouring tissues with different growth rates. When splaying is produced by asymmetrical elongation growth, irregularly formed cells with extensive intercellular systems are created. The spongy parenchyma of foliage leaves is a typical example.

Spacious aerenchymas can also be found in the pith of many monocots living in humid habitats like rush. This tissue type is also called star parenchyma because of its characteristic cell shapes. In the submerged (growing under water) shoots of many angiosperms like the white or yellow water lily exist large air-tubes that can already be seen with the naked eye. They are separated by simple cell layers. This honeycombed construction has two effects: on one hand is a great stability achieved with only little material. This principle has also successfully been applied in technological areas. On the other hand causes the low specific weight a buoyancy that supports the floating of the shoot.

Assimilation Tissue : Assimilation tissues are - in a broader sense - all those tissues that are made from chloroplast-containing cells and are thus able to perform photosynthesis. They are found in all the green parts of a plant. An important aspect of photosynthesis is the integration of carbon dioxide into organic compounds. The resulting products are summed up as assimilates. The leaves of higher plants are by far the most important production centres - if you disregard unicellular aquatic algae.

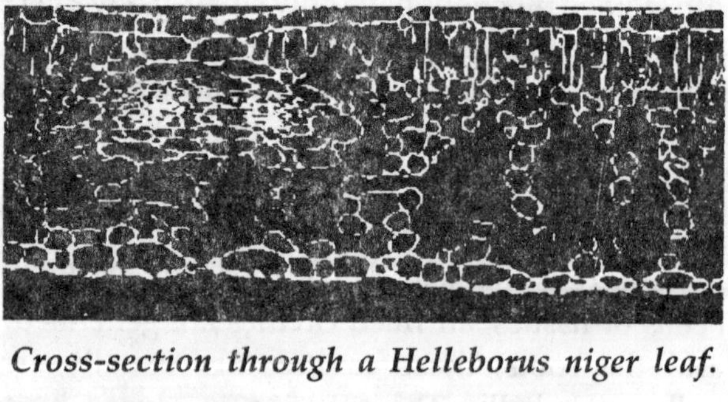

Cross-section through a Helleborus niger leaf.

Leaves consist usually of three different tissues: the mesophyll, the epidermis and the vascular tissues. The mesophyll is a parenchyma tissue. It is a true assimilation tissue. In the leaves of most ferns and phanerogams, it is organised into palisade parenchyma and spongy parenchyma.

Palisade Parenchyma : The cells of the palisade parenchyma are cylindrical. Neighbouring cells look like the stakes of a palisade. The analogy is striking, when looking at the cross-section of a leaf, but is less suitable, when thinking of the three-dimensional arrangement of the cells. Cells of the palisade parenchyma contain three to five times as many chloroplasts as those of the spongy parenchyma. The chloroplasts stay usually near the cell's wall, since this adjustment guarantees optimal use of light. Some species have irregular, or even branched palisade parenchyma cells (Y-shaped: arm palisade cells). Especially ferns, conifers and a few angiosperms (some Ranunculacea- and Caprifoliacea-species like elder) harbour such exceptions. No direct relations between these species exist.

The development and particularly the differentiation of the palisade parenchyma is influenced by extern factors like light and the CO_2-content of the atmosphere. In many species, it is distinguished between sun-and shade leaves. Sun leaves have been exposed to large quantities of light during ontogenesis. This results in a multi-layered palisade parenchyma. Shade leaves in contrast perceive only little light, the palisade parenchyma stays single-layered.

The enlargement of the palisade parenchyma causes usually a reduction of the spongy parenchyma, which is accordingly less well-developed in sun leaves. As important as the exposition to light is the leaf's position at the stem. Old leaves that live near the soil do often have a palisade parenchyma of just one layer's thickness, while younger ones at the top of the plant have normally multi-layered palisade parenchymas.

Spongy Parenchyma : The variability of the cells of the spongy parenchyma and that of its organisation is even greater than that of the palisade parenchyma. It is often said to be an aerenchyma, since it is characterised by a large number of connected intercellular spaces. This does not mean that the contact between palisade parenchyma and intercellular spaces is less well developed. On the contrary: the proportion of palisade parenchyma that is in contact with the intercellular space is larger than that of the spongy parenchyma.

The elongated and cylindrical shape of the cells enables only selective contact to neighbouring cells even at close packing. The spongy parenchyma shows larger contact areas as well as cell-to-cell contacts between palisade parenchyma and spongy parenchyma and between spongy parenchyma and cells of the vascular bundles. This is also the reason for a water and assimilate transport that proceeds without losses.

Supporting Tissues : The development of stable supporting elements has been an important prerequisite for the evolution of large terrestrial organisms. Animals have endo- or exoskeletons that correspond in function to the woody stems or trunks of plants. The architectural design of the plant's body of vegetation is very complex. Thin petioles carry heavy and flat laminas, stems support leaves, flowers and fruits. All plant organs are exposed to mechanical strains. Organs above ground follow the wind's drift. Their high elasticity lets them either return to their original position, or it makes them swing around an imaginative axis.

Trunks are stable enough to resist the wind's pulling. They withstand pressure and are inflexible, although their projecting treetops provide the wind with a large target. The wind makes the upper plant organs and the trunk act like a lever, a large part of the force is hence exerted onto the roots, that anchor the plant in the soil. Other functions of the root are water and nutriment uptake. The strength of tissues protects also against enemies. The hard shell of many seeds prevents a chewing to pieces or puncturing by animals and avoids that parasites like fungi or bacteria force their way into them.

The preceding topic mentioned the high water-content of plant cells that lends a high tension to plant tissues and is caused by the turgor. It supplies plant tissues with a certain stability. Its actual importance is seen best in wilting leaves or flowers after their water supply has been stopped. Extensive specialised supporting tissues exist only in vascular plants. Despite the existence of huge marine brown algae (seaweeds, like Macrocystis, Laminaria), not a single terrestrial alga, whose thallus raises more than a few cell layers above ground, is known. Vascular plants have up to three types of supporting tissue:

1. The collenchyma, a tissue of living cells,
2. The sclerenchyma, a tissue of nearly always dead cells, and
3. The vascular tissue consisting of both living and dead cells. It is responsible for the transport and dispersal of water, nutriments and assimilates.

The larger a vessel plant is, the higher is its content of dead cells. Dead cells are exceptions among bryophytes, but very common in flowering plants. They are usually elongated (prosenchymatous) cells, in parallel to the axis of the respective organ and often combined in sheaves, the fibres. The botanist H.V. MOHL from Tubingen recognised already in the 30th of the 19th century that all these fibres spring from normal, living cells. Supporting tissues reside generally in the periphery of plant organs. If the cells are combined in layers, tubes, whose stability is much greater than that of sticks of the same diameter are formed. The supporting tissues of ribbed or edged stems are concentrated in these ribs or edges. In submerse living vascular plants, the supporting tissue is reduced to a minimum.

Collenchyma : The collenchyma is the typical supporting tissue of the primary plant body and growing plant parts. Its prosenchymatous cells are living at maturity and are always kept in a primary state, which means that they are never lignified. Collenchyma walls are interspersed with groups of pits that tend to be organised in special areas. The name collenchyma derives from the Greek word "kolla", meaning "glue", which refers to the thick, glistening appearance of the walls in fresh tissues. The collenchyma is the typical supporting tissue of the primary plant body and growing plant parts, though it is kept with unaltered structure and function even in outgrown organs like stems, petioles, laminae or roots.

In cross-sections of stems, the collenchyma commonly appears as discrete strands or as a peripheral cylinder that lies, depending on the species, either directly beneath the epidermis or is separated from it by several layers of parenchyma. The cylinder is usually composed of several layers. Collenchyma is also found bordering the veins of dicot leaves. It forms fibres in edgy stems that run along the edges or ribs. Often either phloem or xylem of the vascular bundles is associated with collenchyma cells. Many transitions prove the collenchyma's origin from the parenchyma. The differentiation is reversible, a degeneration to meristematic states has often been observed.

The walls of collenchyma cells are strengthened by the deposit of cellulose and the coating with pectin. These strengthenings are often restricted to single parts or edges of the cell. The walls of parenchyma cells are opened by pits that are often arranged in special areas. The cell walls of collenchyma cells are distortable when stretched. Shape and arrangement of the cells cause a high mechanic stability with a capacity of 10-12 kg/mm^2. This quality is especially advantageous in growing plant organs. It enables the collenchyma cells to stretch in synchrony with the other cells without spoiling the toughness of the tissue. The new state is stabilised by the simultaneous working-in of additional wall material.

Sclerenchyma : The other true supporting tissue is the sclerenchyma. Two groups of sclerenchyma cells exist: fibres and sclereids. Their walls consist of cellulose and/or lignin. Sclerenchyma cells are the principal supporting cells in plant parts that have ceased elongation. Sclerenchyma fibres are of great economical importance, since they constitute the source material for many fabrics (flax, hemp, jute, ramie). Contrary to the collenchyma, mature sclerenchyma is composed of dead cells with extremely thick cell walls (secondary walls) that make up to 90 per cent of the whole cell volume.

The term "sclerenchyma" is derived from the Greek "scleros", meaning "hard". It is their hard, thick walls that make sclerenchyma cells important strengthening and supporting elements in plant parts that have ceased elongation. The difference between fibres and sclereids is not always clear. Transitions do exist, sometimes even within one and the same plant. Fibres are generally long, slender, so-called prosenchymatous cells, usually occurring in strands or bundles. Such bundles or the totality of a stem's bundles are colloquially called fibres. Their high load-bearing capacity and the ease with which they can be processed has since antiquity made them the source material for a number of things, like ropes, fabrics or mattresses.

The fibres of flax (Linum usitatissimum) have been known in Europe and Egypt since more than 3000 years, those of hemp (Canabis sativa) in China for just as long. These fibres, and those of jute (Corchorus capsularis) and ramie (Boehmeria nivea, a nettle), are extremely soft and elastic and are especially well suited for the processing to textiles. Their principal cell wall material is cellulose. Contrasting are hard fibres that are mostly found in monocots. Typical examples are the fibres of many Gramineae, Agaves (sisal: Agave sisalana), lilies (Yucca or Phormium tenax), Musa textilis and others. Their cell walls harbour, besides cellulose, a high proportion of lignin.

The load-bearing capacity of Phormium tenax is as high as 20-25 kg/mm^2 and is thus the same as that of good steel wire (25 kg/ mm^2). But the fibre tears as soon as it is put too great a strain on it, while the wire distorts and tears not before a strain of 80 kg/mm^2. The thickening of a cell wall has been studied in Linum. Starting at the centre of the fibre are the thickening layers of the secondary wall deposited one after the other. Growth at both tips of the cell leads to simultaneous elongation. During development do the layers of secondary material seem like tubes, of which the outer one is always longer and older than the next. After completion of growth the missing parts are supplemented, so that the wall is evenly thickened up to the tips of the fibres.

Fibres stem usually from meristematic tissues. Cambium and procambium are their main centres of production. They are often associated with the xylem of the vascular bundles. The fibres of the xylem are always lignified. Reliable evidence for the fibre cells' evolutionary origin of tracheids exists. During evolution the strength of the cell walls was enhanced, the ability to conduct water was lost and the size of the pits reduced. Fibres that do not belong to the xylem are bast (outside the ring of cambium) and such fibres that are arranged in characteristic patterns at different sites of the shoot.

Sclereids are variable in shape. The cells can be isodiametric, prosenchymatic, forked or fantastically branched. They can be grouped into bundles, can form complete tubes located at the periphery or can occur as single cells or small groups of cells within parenchyma tissues. But compared with most fibres sclereids are relatively short. Characteristic examples are the stone cells (called stone cells because of their hardness) of pears (Pyrus communis) and quinces (Cydonia oblonga) and those of the shoot of the wax plant (Hoya carnosa). The cell walls fill nearly all the cell's volume. A layering of the walls and the existence of branched pits is clearly visible. Branched pits such as these are called ramiform pits. The shell of many seeds like those of nuts as well as the stones of drupes like cherries or plums are made up from sclereids.

Vascular Tissues : Vascular tissues do not only fulfil supporting functions. Their conductive functions are more important. They consist of water-conducting xylem and food-conducting phloem. Only the xylem still supporting functions. Xylem and phloem are combined in a structure that is called vascular bundle. Vascular tissues transport water, dissolved salts, products of photosynthesis (assimilates), growth regulators like phytohormones and, last but not least, harmful substances and parasites like viruses or mycoplasmas. It is commonly distinguished between the elements of water conduct, the xylem, and those of assimilate conduct, the phloem. Both are combined in vascular bundles.

Vascular bundles contain mostly dead cells. In contrast to all other tissues of the actual topic, the phloem, that will be reviewed here together with the xylem has no supporting function whatsoever. The lignified cell walls of the xylem elements will at the moment, rather simplifyingly, be called wood. They can last for thousands of years. Their petrified states are ideal evidences for the first occurrence and the successive spreading of the great taxonomical groups of terrestrial plants.

Genuine vascular bundles are the name-giving feature of vascular plants (Trachaeophtes: Pteridophytes and Phanerogames). Thallophytes have at most first signs of a conductive system, which is not homologous to the respective tissues in vascular plants. Primitive variations of conductive elements appear also in mosses. The occurring cell types and the arrangement of the vascular bundles within shoots and roots are reliable features for the characterisation of individual classes of plants. The xylem of nearly all angiosperms contains three types of cells:

- Wood vessels
- Tracheids and
- The already mentioned xylem fibres.

Most gymnosperms lack wood vessels. The few exceptions point out the relationship between gymnosperms and angiosperms. The vascular bundles of monocots seem to run through the shoot without any obvious order, while those of dicots and many gymnosperms are arranged in peripheral rings. Xylem and phloem are separated by a meristematic tissue, the vascular cambium. It provides the means for secondary growth. The vascular bundles of pteridophytes and the roots of flowering plants are located centrally.

Xylem : The xylem is the principal water-conducting tissue of vascular plants. It consists of tracheary elements, tracheids and wood vessels and of additional xylem fibres. All of them are elongated cells with secondary cell walls that lack protoplasts at maturity. Bordered pits are typical for tracheids, while wood vessels are marked by perforated or completely dissolved final walls. The xylem takes also part in food storage, support and the conduction of minerals. Xylem and phloem together form a continuous system of vascular tissue extending throughout the plant.

The principal conductive cells of the xylem are tracheary elements, of which there are two types, tracheids and wood vessels. Both are elongated cells with secondary cell walls that lack protoplasts at maturity. They are completed by the xylem fibres and parenchyma cells. Much speaks on behalf of the origin of xylem fibres and wood vessels from the tracheids. Since 1851, the isolation and depiction of lignified cells is done according to the maceration procedure of Schulze. Small pieces of wood are covered with a mixture of potassium perchlorate and concentrated nitric acid.

The complete volume should not be larger than 1/10 of the reaction container, because a lot of gas develops very quickly at cautious heating. The surfaces of the wooden pieces are strongly attacked. The single cells can be scraped off after washing of the preparation and examined under the microscope. Let us start with a portrait of the different cell types. Tracheids are the chief water-conducting elements in gymnosperms and seedless vascular plants. They can also be found in angiosperms. Tracheids are elongated cells, closed at both ends. They are 1 mm on average.

Tracheids are regarded as the prototype of prosenchymatic cells, since the cell's ends are pointed and true final walls are missing. Tracheids look often square in cross-section, the lignified secondary wall is relatively thin. Their entire cell surface is evenly coated. The walls are opened by numerous pits that are, depending on their origin, either round, oval, gap- or groove-shaped. They occur solitarily, statistically scattered, arranged in turns around the middle axis or grouped together. Such groups can often be found at the cell's ends. If gap-like pits lay on top of one another, a ladder- or stair-like perforation may be the result. It is commonly called scalariform.

The pits are often surrounded by a halo and are then called bordered pits. Bordered pits are especially common in the tracheids of some gymnosperms. Their structure can be discerned best in a cross-section through neighbouring cells. The middle lamina between the cells is preserved within the pits. Their centre is made up by a disc of primary cell wall material, called torus. No secondary walls exists in the pit's structure. The area between torus and wall (the former middle lamina) is called margo and is very porous, allowing the movement of water and ions from tracheid to tracheid. Bordered pits exist only in cells with secondary walls.

Botanists think of wood vessels (tracheae) as the water-filled tubes of the xylem. M. Malpighi, who thought that he had found an important common element in the anatomy of animals and plants introduced the term trachea in the 17th century. Wood vessels are the chief water-conducting elements of angiosperms. In contrast to the tracheids the final walls of the single vessels are perforated or, much more so, completely resolved. Wood vessels are therefore generally thought to be more efficient water conductors than tracheids.

The length of the single tube (composed of numerous cells) makes it difficult to isolate a vessel as a whole. It can be as long as several metres. It is commonly assumed that at least in some species the wood vessels are as long as the whole shoot. During ontogenesis the wood vessels increase strongly in width. They are usually round in cross-section and have a larger diameter than the tracheids, a feature that enhances their capacity for water-conduct. Exceptionally wide-lumened elements can be seen with deciduous trees, that are known to lose particularly large amounts of water due to transpiration. The water-loss of a fully developed birch tree with an estimated number of 200 000 leaves can be up to 400 litres per day.

Even wider are the vessel elements of many lianas. But the

oldest living trees, the redwoods and other sequoias at the pacific coast of California have without exception tracheids with very narrow lumina. Vessels are marked by characteristically structured secondary wall coatings (lignin) at the inner surface of the primary walls. Deposits in the form of screws, rings or nets exist. These strengthenings make it possible for tracheary elements to be stretched or extended, although the cells are frequently destroyed during the overall elongation of the organ.

Beside this vessels with pits or scalariform openings exist, whose walls are nearly completely lined with secondary wall material that is only opened by round or gap-like pits. Many transitions between the two pit types can be found. Often some or even all of the types are members of the same vascular bundle. But there are species, that lack one or the other type. Wood vessels develop -just like tracheids- during primary growth from the cells of the procambium. Where secondary growth occurs wood vessels are produced by the cells of the cambium.

Phloem : The other vascular tissue is the phloem. Its principal function is the conducting of assimilates and food. It is composed of the sieve elements, of which two types can be distinguished, sieve cells and sieve-tube members. Phloem elements do typically have sieve plates instead of final walls. Sieve-tube members of angiosperms are associated by living companion cells. The phloem is the principal food-conducting tissue of vascular plants. Its elements are elongated, just like those of the xylem. In contrast to tracheids and wood vessels, mature phloem elements contain a protoplast and sometimes even a nucleus.

Phloem elements may be of primary or secondary origin though the early primary phloem, the protophloem, is frequently destroyed during elongation of the respective organ. The main conducting elements of the phloem are the sieve elements, of which there are two different types: sieve cells and sieve-tube members. Sieve cells have narrow pores, their sieve areas are quite uniform in structure, and they are distributed evenly on all walls. One of the principal differences between sieve cells and sieve-tube members is the presence of sieve plates in sieve-tube members, that are absent in sieve cells.

Sieve cells are the only type of food-conducting cells in most seedless vascular plants and gymnosperms, whereas in angiosperms only sieve-tube members are present. Sieve-tube members occur end-on-end in longitudinal series called sieve tubes. They are in contact via plasmodesmata. Typically, the final walls are interspersed with primary pit areas (groups of plasmodesmata), that later on develop into sieve plates. Sieve tubes in the phloem of angiosperms are flanked by one or several plasma-rich, nucleated companion cells, that do not occur in gymnosperms. Much less is known about the phylogeny of sieve elements than about that of xylem members.

One cause are the cell walls of sieve elements, that are strengthened entirely by cellulose and are therefore not as resistant as lignified ones, another is the transitory working order of the single sieve element. After loss of function the cells are reorganised and lose their typical structural properties. Good fossils are thus rare. Phloem elements of the roots are normally organised in a more progressive and thus more efficient way than their equivalents in the shoots. Callose is deposited in the sieve plates at regular intervals and can be detected with special reagents. Resorcin blue gives a blue, aniline blue a bright yellow staining.

The amount of callose increases with cell age, continuously reducing the diameter of the pores. Sieve elements that have lost their function are blocked by thick plugs of callose. The question, what is cause and what result remains. Are the cells inactivated by the callose plugs or do these form as a result of the loss of function? Active sieve tubes contain huge amounts of so-called P-proteins (phloem - proteins). It has often been speculated, whether they have an active part in assimilate transport. No answer has been found up until now.

The sieve tube elements of angiosperms (mono- and dicots), but not those of pteridophytes and gymnosperms are associated with companion cells. In Ginkgo and other gymnosperms their function is taken over by specialised parenchyma cells, that resemble companion cells in structure: the albuminous cells. They work out the contacts between phloem and the surrounding transfusion parenchyma and can also be found as mediators between ray parenchyma cells of the bark and mature sieve cells. Albuminous cells participate in loading and unloading of these cells. The process has been studied in detail in Pinus.

Companion cells and sieve elements are of meristematic origin, their development is similar to that of the xylem. Primary phloem develops by longitudinal division and subsequent elongation of meristematic cells. The cells do often divide unequally. The bigger daughter cell differentiates into a sieve element, the smaller after one or two further longitudinal divisions into two to four companion cells. Consequently several companion cells may belong to one sieve cell. The exact number is usually typical for the respective tissue, but it may even vary within a single plant. Beside the typical phloem elements, fibres and/or sclereids as well as parenchyma cells that serve as deposits for starch, fat, oil and other nutrients, can belong to the phloem. The deposits contain often tannic acids and resins.

Vascular Bundles : Different types of vascular bundles evolved in the different plant groups. Monocots have usually collateral vascular bundles, a type that is also called closed. The vascular bundle is enclosed by a bundle sheath of parenchyma cells. Xylem and phloem of the shoots of mono- and dicots are usually arranged opposite to each other in the vascular bundles (collateral vascular bundles). The xylem is normally at the inside, the phloem at the outside. In dicots, both components are separated by the fascicular cambium. It is also spoken of an open vascular bundle in contrast to the 'closed' bundles of monocots that lack the cambium.

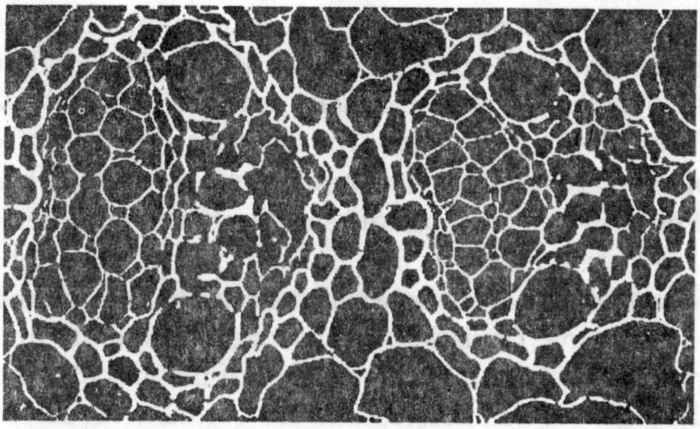

Xylem and phloem are surrounded by a bundle sheath of parenchyma that is often starch-containing. The prime object for the demonstration of structure and arrangement of monocot bundles is the shoot of corn (Zea mays). Although the vascular bundles seem to be scattered in cross-section, do they occur at the periphery in larger numbers. Central bundles are usually bigger than peripheral ones. Many monocots have hollow stems. It seems as if their vascular bundles were arranged in a ring, because space is rare. In the stems of some aquatic monocots (Elodea, Potamogeton), they are united in a central (axial) strand. This arrangement is of secondary nature, since it developed as an adjustment to aquatic life.

A comparable structure can be found in some submerse dicots. It decreases the strength of the stem, while at the same time increasing its flexibility. Exactly this property is needed to meet the demands of running water. Selection favouring efficient water transport cannot be expected in submerse living plants anyway.

The collateral open vascular bundles of dicots display xylem and phloem that are separated from each other by the fascicular cambium. These vascular bundles are also often surrounded by a bundle sheath and the whole structure is imbedded in a parenchyma tissue called pith. Another type of vascular bundles are the bicollateral bundles.

The circular arrangement of vascular bundles can be

observed very well in cross-sections of the stem. Interfascicular cambium occurs in many, but not all species. Herbaceous dicots do always have a fascicular cambium, that is, however, usually not divided any more. The structural elements of xylem and phloem are the same as in monocots. The vascular bundles are often enclosed by a bundle sheath of cells with thickened walls and are embedded in a parenchyma of large cells, the pith. Parenchyma areas between the bundles are often called ray parenchyma. Deviations of this organisation are found in square stems, where the arrangement of the vascular bundles and the shape of the cross-section are in tune with each other.

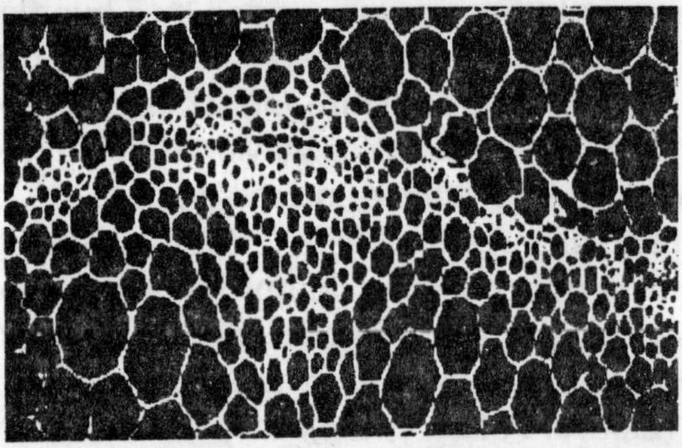

A largely changed pattern appears in some families (Solanaceae, Asclepidaceae, Compositae, Curcubitaceae and others), whose vascular bundles are open and bicollateral, i.e. the xylem is on both sides associated with phloem. The vascular bundles are arranged in two concentric circles, the lumina of the inner circle being much wider than those of the outer. Aristolochia sipho has proven useful for the demonstration of secondary growth and thus also for the development of wood. Very young branches have merely any fascicular cambium. Interfascicular cambium develops very early during the first year even before the beginning of secondary growth. This pattern is largely kept in branches of several years' age.

The activity of the cambium increases their diameter and the vascular bundles become elongated in cross-section. In general, far more xylem than phloem elements are produced. Annual rings are clearly visible, because at the beginning of each vegetation period (in spring), vessels (conducting function) and fibres (supporting function) with a wide lumen are assembled first, the so-called early wood. In the following, elements with steadily narrowing volumes are produced. In autumn, only few vascular elements with narrow lumina (late wood) form.

Secondary Growth

Gymnosperm wood : Gymnosperm wood is more homogeneous than that of angiosperms. Its only conducting elements are tracheids. Just like the wood of angiosperms, it is interspersed with radially orientated rays (medullary rays) consisting of parenchyma cells and sometimes also tracheids. The wood of gymnosperms is simpler and more homogeneous than that of angiosperms. Except for the species of the order Gnetales, tracheids are the only conducting elements in gymnosperms.

The processes within angiosperm wood are in principal the same, but it was at first by no means simple to recognise the logical connection. Pinus silvestris or related species like the North American white pine (Pinus strobus) turned out to be ideal specimens. Pine wood is a typical object of every botanical ground course. Since cells and tissues are three-dimensional objects, they do have to be regarded from three sides to get an impression of their spatial organisation:

- Cross-section
- Radial longitudinal section
- Tangential section.

This maybe best-known property of gymnosperm- and angiosperm wood beside the annual rings is the grain, that is visible especially nicely in tangential sections. The veneers often used in the fabrication of furniture are made from such sections. The xylem of gymnosperms has few or no parenchyma cells at all. Their existence (or non-existence) is a feature of certain genera. With Pinus, they are found only in the epithelia of resin ducts, with many Podocarpaceae, Taxodiaceae and Cupressaceae (cypresses), they are amply present, while they are missing completely with Araucariaceae and Taxaceae (yews).

The tracheids of gymnosperms are 0.5-11 mm long and are orientated along the shoot or root axis. They border at neighbouring tracheids above and below not with their final walls, but with the ends of their lateral walls. This is the reason, why there are never, neither in tracheids nor in vessels, ideal, vertical conducting tubes. In many species, it is distinguished between sapwood and heartwood. Sapwood is an active, water-conducting tissue, while heartwood is inactive and has only supporting functions. Its cells contain but little water or reserve compounds. Instead, organic compounds, oils, rubber, resins, tannic acids, dyes or aromatic compounds are stored here.

Oxidised phenolic compounds give the wood a dark colour. The typical heartwood is missing in spruce (Picea excelsa), pine (Abies alba) and in some angiosperms (poplar, willow). These woods are regarded as less valuable by the timber industry than heartwood-containing ones. Medullary rays Wood is at regular distances interspersed with radially orientated parenchyma cells and often also with tracheids. They spring from the ray initials> of the cambium. The medullary rays of conifer wood are normally only one cell layer thick, but they can be as high as 1-20 (sometimes even up to 50) layers of cells.

Ray tracheids and axially orientated tracheids are connected via pits. Medullary rays with resin ducts seem spindly in tangential sections. Resin ducts can be either axially or radially orientated. There is actually hardly any difference between them and intercellular spaces that have been enlarged by the splaying of parenchyma cells. Consequently, resin ducts are always lined with parenchyma cells. Their existence has different reasons: injuries by frost or winter damages are some factors that stimulate their formation. The respective gymnosperm families react differently to these disturbances.

Angiosperm wood : The structure of angiosperm wood is more inhomogenous and variable than that of gymnosperms. It can be of two different vessel arrangement patterns: it belongs either to wood with statistically scattered pores or to wood with ring-shaped pores. Angiosperm wood does also contain rays. They are often considerably larger than those of gymnosperms. Angiosperm wood in its typical structure. is found almost only in dicots. First signs of wood formation occur in some monocot species, but their wood is always built inhomogenously.

Contrary to the very clear architecture of gymnosperm wood, that of angiosperms is more complex and variable, partly because of a greater number of different cell types in the axial system. Most wood contains vessel members as conducting elements and they can be structured very differently. Over and above the wood may hold a high proportion of xylem fibres and parenchyma cells (the latter constitutes up to 23 per cent of all xylem elements). The organisation of rays varies also, they include almost only parenchyma cells, but parenchyma of strongly varied shaping. The differences in shape and orientation of xylem elements can be explained by the organisation of the cambium.

Many tropical species develop no annual rings. 75 per cent of the trees of the Indian rain forest, 45 per cent of those of the Amazonas Basin and 15 per cent of the Malaysian trees have no annual rings. Plants of dry habitats, that are in permanent contact with the ground water like Tamarix articulata, Acacia raddiana or Acacia tortilis show steady activity throughout the year. Transplantation experiments revealed that the rhythm of cambium activity is inherited in some species, while it is chiefly controlled by extern factors in others.

The vessels are connected laterally (exchange of solutes via pits). In some species, the contact is restricted to vessels in growth zones, while it extends over several vessels in others. Ample xylem fibres are common in most angiosperm woods.

In wood with ring-shaped pores, they are predominant in late wood. In some species, the xylem fibres contain living protoplasts that serve generally as starch deposits. This type of organisation is estimated progressive. Lacking or small numbers of axial parenchyma cells are strikingly correlated with this type of organisation.

Rays of angiosperm wood contain only rarely tracheids, but are constructed from parenchyma cells, whose longitudinal axis is either orientated radially or axially. Both single layered and multi-layered rays occur. The rays of dicots are often considerably larger than those of gymnosperm wood. They range from one to many cells wide and from one to several hundred cells high. Just like the wood of gymnosperms, that of angiosperms is interspersed with numerous intercellular spaces, whose structure, development and content are, however, more variable.

Besides channels, cavities containing resins, oils, rubber or mucigels occur often. Their walls may or may not be lined with parenchyma cells. The intercellular spaces are, just like in gymnosperms, formed by splaying or ripping. Rubber-like substances are often secreted only after degeneration of the cells. This may also lead to filling and subsequent loss of function of neighbouring cells. The production of rubber can be caused by infections, damage due to insects or physiological disorders. Structure and arrangement of xylem elements determines the strength and other physical properties of the wood and thus its economic value.

An important, though not exclusive, indication of the strength is the specific weight. The specific weight of dry, water-free wood is dependent on the mass of the cell wall material per unit volume. The weight of pure wall substance is 1.4-1.62 g/cm^3; the same substance measured within the intact tissue has a weight of 0.04-1.4 g/cm^3. The value 0.04 was measured for the wood of a leguminosae (Aeschynomene), while the value 1.4 has been found in a species of the Rhamnaceae family (Krugiodendron). The difference is founded on the way the tissues are constructed and on the relative amounts of the individual xylem elements.

Xylem fibres with thick walls and small lumina add to the increase of the specific weight, wide lumina and thin walls decrease it. On the other hand, the strength is amplified by xylem fibres and weakened by vessel elements. Wood with ring-shaped lumina is less resistant to some strains due to the concentration of the vessels in defined areas than wood with diffusely dispersed vessels. The famous and astonishingly stable balsa wood (Ochroma pyramidale, a species of the Bombacaceae family) has a specific weight of 0.1-0.16 g/cm^2. This low weight is partly caused by the high proportion of large, thin-walled parenchyma cells. Two general principles of construction underlie light wood. Either layers of thick- and thinwalled cells alternate or both cell types are evenly distributed and form a structurally homogeneous tissue.

Monocotyledon Wood : The structure of monocot wood is fundamentally different from that of dicots. Vascular bundles strengthened by extremely tough fibres with small lumina are embedded into voluminous parenchyma that consists of regularly arranged cells with large lumina. The wood of monocots is built on a 'reinforced concrete principle' that lends it a great strength of flexibility. Just think of bamboo or the high, slender palms that resist even the strongest tropical storms, or the liana-like rattan cane, whose wood is used for the production of rattan furniture. Gymnosperm wood, however, is constructed on the 'brick principle'. It is thus not astonishing that the wood of spruce or pine splinters easily.

Vascular Bundles of Roots : The vascular bundles of roots are enclosed by two concentric tubes of cells, the pericycle and the endodermis. The whole complex is embedded in the primary bark. Casparian strips are a typical feature of the vascular bundles of roots. Roots are able to perform secondary growth, which starts with the development of secondary cambium and leads to a restructuring of the vascular bundles. A feature of roots with secondary growth are feeder roots. The vascular bundles of roots are constructed rather simply and show but little variations, if compared to those of the shoot. One reason could be the less variable surrounding that poses fewer demands than the surrounding above ground.

Xylem- and phloem elements are combined in a central strand without pith, the stele. The root of dicots is often regarded as the basic type. The root of monocots differs only negligibly as far as the primary construction is concerned so that a separate reviewing is unnecessary. The roots of many dicot species like wood plants or perennial plants are able to perform secondary growth, while roots of monocots can, with few exceptions like Dracaena, not grown in this way. The primary cylinder of vascular bundles is as a rule composed of radially located xylem elements and peripheral phloem.

The vascular elements are enclosed by two concentric tubes of parenchyma tissue, the pericycle (inner cylinder) and the endodermis (outer cylinder). The pericycle is missing in the roots of some simple aquatic plants and in some parasitic species. The whole complex is embedded into a parenchyma of large cells the primary bark. The relation of the vascular bundle arrangements of roots and shoots is illustrated by the theory of steles. It is based upon the assumption that root and shoot are merely two parts of one axial unity.

A continuously recurring problem is the question of the vascular bundles' reorganisation in the zone of transition between both organs. It should, strictly spoken, not be talked of a root to shoot transition, since the two organs are separated by both epicotyl and hypocotyl. These are formed first in the development of the embryo and only later are shoot and roots laid out. During this early phase, vascular bundles are formed in the area of the epi- or hypocotyl that elongate, beginning at the point of initiation, simultaneously both in the direction of the shoot and the roots and in this process adopt the organ-specific arrangement.

Secondary growth occurs, as has already been mentioned

above, in the roots of many dicots. In the course of this, as a rule a secondary cambium develops out of parenchyma cells that have been laid out during primary growth between the xylem rays. The cambium is supplemented by cells of the pericycle so that finally a homogeneous cambium cylinder is formed. As a result of the secondary growth, pericycle and epidermis tear apart. The consequent gaps are filled by division of additional parenchyma cells. Just like shoots, roots develop a secondary final tissue (phelloderm, phellogen, phellem).

Although the vascular bundles of roots and shoots share many principles of organisation, some fundamental differences exist. The formation of the shoot's vascular bundles is normally linked to the simultaneous lay out of leaves or branches and starts at the apical meristem. Lateral roots spring from cells of the pericycle and grow from the inside towards the outside. Only in a second, independent step of differentiation is the contact of the newly developed vascular bundles to the primary (tap) root established. The lateral roots may branch again. In this way, side roots of 2., 3.,.... order develop.

Contrary to leaves, side roots are built with central symmetry in a strictly morphological sense (exceptions are, for example, the roots of Kalanchoe), while leaves and leaf stalks are normally organised dorsoventrally. But a physiological dorsiventrality seems also to exist in lateral roots. It is known that roots perform two in some respects contradictory tasks. On one hand they anchor the root in the soil, on the other hand they take up water and nutrients from the soil. The anchor function demands a tension-proof root, interspersed with strengthening elements like lignification or fibres within the phloem. This function is supported by secondary growth.

On the other hand lignification and formation of bark decreases the ability for water uptake and transport. Therefore has, also in roots, the division-of-labour-principle' been put through. Perennial plants in particular have at the endings of strongly lignified lateral or primary roots unlignified, short-lived branchings, called feeder roots. Additionally the secondary final tissues are not completely water-tight. How far they are engaged in the uptake of nutrients is not yet known. Beside the primary and lateral roots, many mono- and dicots are equipped with so-called adventitious roots. They originate in different parts of the shoot or root. The root system of many Poaceae, for example, consists nearly entirely of adventitious roots. Often are roots or parts of them including the hypocotyl used as storage organs. The parenchyma tissue of the primary bark, the xylem and the phloem swell up and numerous cell divisions take place. The newly formed deposits are secondarily supplied with vascular bundles.

Animal Cells and Tissues

Organs in animals are composed of a number of different tissue types. For example, the stomach:

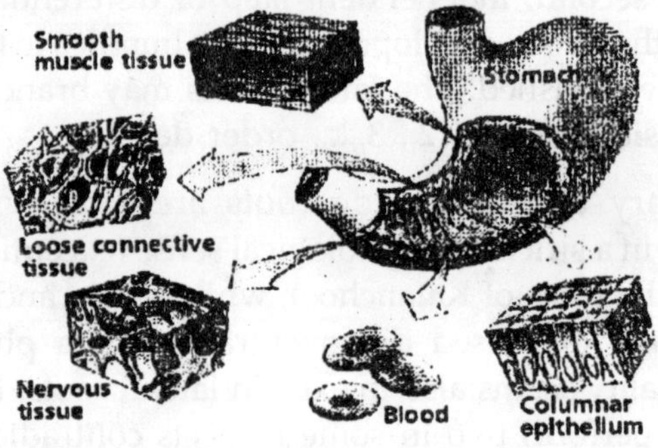

Cells and tissues that comprise the stomach.

Plants are simpler organisms than animals, having three organ systems and fewer organs than do vertebrate animals. Organs are composed of tissues, which are in turn composed of cells. Plants have three tissue types: ground, demal, and vascular. Animals have four: epithelial, connective, muscle, and bone.

Epithelial Tissue : Epithelial tissue covers body surfaces and lines body cavities. Functions include lining, protecting, and forming glands. Three types of epithelium occur:

- Squamous epithelium is flattened cells.
- Cuboidal epithelium is cube-shaped cells.
- Columnar epithelium consists of elongated cells.

Any epithelium can be simple or stratified. Simple epithelium has only a single cell layer. Stratified epithelium has more than one layer of cells. Pseudostratified epithelium is a single layer of cells so shaped that they appear at first glance to form two layers.

Functions of epithelial cells include:

- Movement materials in, out, or around the body.
- Protection of the internal environment against the external environment.
- Secretion of a product.

Glands can be single epithelial cells, such as the goblet cells that line the intestine. Multicellular glands include the endocrine glands. Many animals have their skin composed of epithelium. Vertebrates have keratin in their skin cells to reduce water loss. Many other animals secrete mucus or other materials from their skin, such as earthworms do.

Connective Tissue : Connective tissue serves many purposes in the body:

- Binding
- Supporting
- Protecting
- Forming blood

- Storing fats
- Filling space

Connective cells are separated from one another by a non-cellular matrix. The matrix may be solid (as in bone), soft (as in loose connective tissue), or liquid (as in blood). Two types of connective tissue are Loose Connective Tissue (LCT) and Fibrous Connective Tissue (FCT). Fibroblasts (LCT) are separated by a collagen fibre-containing matrix. Collagen fibres provide elasticity and flexibility.

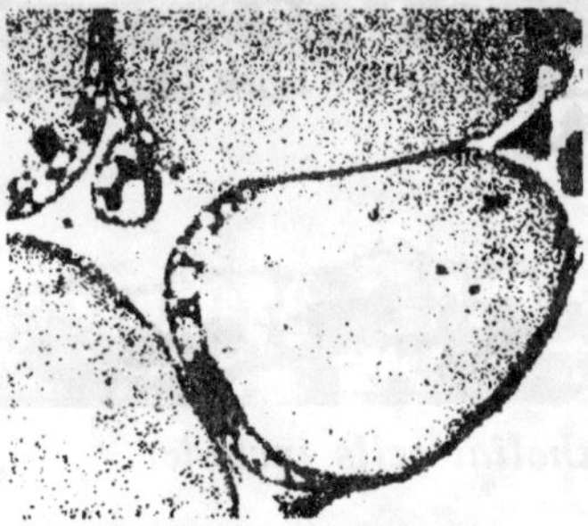

Adipose tissue

LCT occurs beneath epithelium in skin and many internal organs, such as lungs, arteries and the urinary bladder. This tissue type also forms a protective layer over muscle, nerves, and blood vessels.

Adipose tissue has enlarged fibroblasts storing fats and reduced intracellular matrix. Adipose tissue facilitates energy storage and insulation. Fibrous Connective Tissue has many fibres of collagen closely packed together. FCT occurs in tendons, which connect muscle to bone. Ligaments are also composed of FCT and connect bone to bone at a joint. Cartilage and bone are "rigid" connective tissues. Cartilage has structural proteins deposited in the matrix between cells. Cartilage is the softer of the two. Cartilage forms the embryonic skeleton of vertebrates and the adult skeleton of sharks and rays. It also occurs in the human body in the ears, tip of the nose, and at joints.

Bone has calcium salts in the matrix, giving it greater strength. Bone also serves as a reservoir (or sink) for calcium. Protein fibres provide elasticity while minerals provide elasticity. Two types of bone occur. Dense bone has osteocytes (bone cells) located in lacunae connected by canaliculi. Lacunae are commonly referred to as Haversian canals. Spongy bone occurs at the ends of bones and has bony bars and plates separated by irregular spaces. The solid portions of spongy bone pick up stress.

Blood is a connective tissue of cells separated by a liquid (plasma) matrix. Two types of cells occur. Red blood cells (erythrocytes) carry oxygen. White blood cells (leukocytes) function in the immune system.

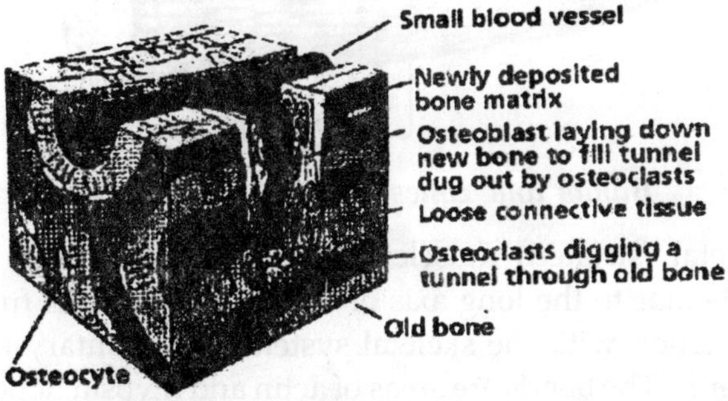

Structure and vascularisation of bone.

Plasma transports dissolved glucose, wastes, carbon dioxide and hormones, as well as regulating the water balance for the blood cells. Platelets are cell fragments that function in blood clotting.

Muscle Tissue : Muscle tissue facilitates movement of the animal by contraction of individual muscle cells (referred to as muscle fibres). Three types of muscle fibres occur in animals (the only taxonomic kingdom to have muscle cells):

- Skeletal (striated)
- Smooth
- Cardiac

Muscle fibres are multinucleated, with the nuclei located just under the plasma membrane. Most of the cell is occupied by striated, thread-like myofibrils. Within each myofibril there are dense Z lines. A sarcomere (or muscle functional unit) extends from Z line to Z line. Each sarcomere has thick and thin filaments. The thick filaments are made of myosin and occupy the centre of each sarcomere. Thin filaments are made of actin and anchor to the Z line.

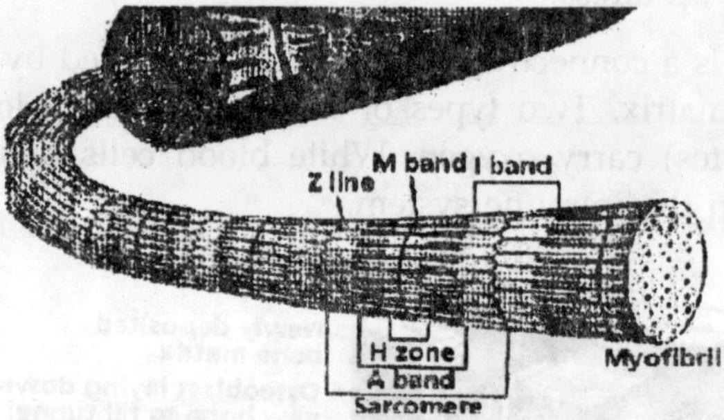

Bands and zones in a muscle fibre.

Skeletal (striated) muscle fibres have alternating bands perpendicular to the long axis of the cell. These cells function in conjunction with the skeletal system for voluntary muscle movements. The bands are areas of actin and myosin deposition in the cells.

Smooth muscle fibres lack the banding, although actin and myosin still occur. These cells function in involuntary movements and/or autonomic responses (such as breathing, secretion, ejaculation, birth, and certain reflexes). Smooth muscle fibres are spindle shaped cells that form masses. These fibres are components of structures in the digestive system, reproductive tract, and blood vessels.

Cardiac muscle fibres are a type of striated muscle found only in the heart. The cell has a bifurcated (or forked) shape, usually with the nucleus near the centre of the cell. The cells are usually connected to each other by intercalated disks.

Nervous Tissue : Nervous tissue functions in the integration of stimulus and control of response to that stimulus. Nerve cells are called neurons. Each neuron has a cell body, an axon, and many dendrites. Nervous tissue is composed of two main cell types: neurons and glial cells. Neurons transmit nerve messages. Glial cells are in direct contact with neurons and often surround them. The neuron is the functional unit of the nervous system. Humans have about 100 billion neurons in their brain alone! While variable in size and shape, all neurons have three parts. Dendrites receive information from another cell and transmit the message to the cell body.

The cell body contains the nucleus, mitochondria and other organelles typical of eukaryotic cells. The axon conducts messages away from the cell body.

of stimulus and control of response to that stimulus. Nervous cells are called neurons. Each neuron has a cell body, an axon and many dendrites. Nervous tissue is composed of two main cell types: neurons and glial cells. Neurons transmit nerve messages. Glial cells are in direct contact with neurons and often surround them. The neuron is the functional unit of the nervous system. Humans have about 100 billion neurons in their brain alone! While variable in size and shape, all neurons have three parts. Dendrites receive information from another cell and transmit the message to the cell body.

The cell body contains the nucleus, mitochondria and other organelles typical of eukaryotic cells. The axon conducts messages away from the cell body.

7

Placental Development

Growth hormone (GH), prolactin and leptin all have a critical and potentially interactive role in regulating placental and fetal development. Plasma concentrations of each may be significantly compromised throughout pregnancy, thereby contributing to altered placental function or fetal growth and development. Each hormone has a primary role in the regulation of both energy balance and tissue development. Perturbations in nutrient supply in conjunction with reduced cytokine receptor abundance can therefore have a profound effect on fetal development, which may determine survival after birth and subsequent morbidity and mortality.

GH is widely recognised to regulate post-natal growth, but there is an increasing amount of evidence demonstrating that GH can enhance embryonic growth. Studies in mice have shown GH and prolactin receptors to be present in mouse zygotes, cleaving embryos and blastocysts. Although both receptor types are present in the inner cell mass and trophectoderm of blastocysts, receptor abundance is much greater in the trophectoderm, indicating a specific role in early placental development. This contrasts with insulin and insulin-like growth factor (IGF)-I, which only have a anabolic effect on the inner cell mass, an effect proposed to be mediated by inhibition of apoptosis. In vitro studies have shown that both GH and prolactin can have an anabolic effect, producing blastocysts with increased cell numbers. It is therefore likely that alterations in maternal GH and/or prolactin secretion will significantly impact on placental growth.

Prolactin receptor deficiency is also associated with reproductive failure. This can be partially overcome by treatment with progesterone, although the number of pups surviving decreases from mid pregnancy and those surviving to term is 3-fold lower than for wildtype controls. Whether this response may be mediated through impaired function of the placenta, in which prolactin receptors are highly abundant, is unknown. The extent to which leptin modulates placental or fetal growth remains a matter of debate. Leptin may only have an indirect effect on reproduction, acting through enhancement of GH secretion. Leptin is critical for the onset of reproductive activity, but not for implantation or subsequent maintenance of pregnancy, in ob/ob mice at least". Direct effects on placental or fetal growth have yet to be documented, despite the finding that leptin is present in the placenta and is influenced by nutritional state.

Placental Development

It is not only gross changes in placental sise that contribute to a large or small fetus. Longitudinal studies from the Dutch famine have shown that, when maternal energy intake was reduced to close to 50 per cent of pre- or post-famine levels, marked changes in placental and fetal weight ensued. Surprisingly, if mothers were only subjected to the famine in early gestation, at term they had a larger placenta and longer infants compared with mothers who were nutrient-restricted in mid or late gestation.

As adults, these infants have been shown to exhibit a more

atherogenic lipid profile, to be more obese and to be at increased risk of coronary heart disease. It is likely that the apparent increase in placental growth is programmed by the altered maternal endocrine environment in early gestation. GH secretion has been established to increase in response to under nutrition. It is hypothesised that increased exposure of the developing embryo to GH could have a direct effect on placental growth.

Although GH does not appear to have a direct role in fetal growth, when its secretion is markedly enhanced and adult-type GH receptors are not present GH may be able to influence subsequent fetal growth by initial effects on placentation.

Cytokines and fetal growth : Although in human fetuses the plasma leptin concentration increases with gestational age, there is a substantial increase, both within and between studies, in the range of values obtained as term approaches. The reason for this remains in doubt, but it may be partly explained by the relative insensitivity of some of the commercially available assay kits that have been used. There is also an appreciable decrease in plasma leptin concentration within the first few hours of birth, which may commence in utero, thereby contributing to the wide variations in published results.

Furthermore, as the rise in plasma leptin secretion with gestation may be of fetal (i.e. adipose tissue) rather than placental origin, differences in adipose tissue weight at birth may contribute to differences between groups. It is likely, however, that a close relationship exists between birth weight and plasma leptin concentration, since in sheep (a species which has very limited adipose tissue stores at birth) a strong correlation between plasma leptin and birth weight was shown when samples were taken hourly over the first 6 h of neo-natal life ($r^2 = 0.94$).

Prolactin has been shown to be present in the fetal circulation in several species, but it does not cross the placenta from ewe to fetus. In the fetal sheep, prolactin synthesis and secretion are both highly regulated prior to birth. The role of prolactin in the growth and development of the fetus, however, remains poorly understood. Prolactinm RNA levels in the fetal pituitary gland and the plasma prolactin concentration are maintained by a tonic stimulatory drive from the hypothalamus, and rise gradually during late gestation. Prolactin receptors are present in a large number of fetal tissues in all species studied to date.

For example, mRNAs for both the long and short forms of the prolactin receptor are present in the fetal sheep liver, and a gestational-age-related increase in the level of receptor mRNA occurs in parallel with the increase in plasma prolactin concentrations. The stimulus for this increase does not appear to be mediated directly by cortisol, but may be an indirect response acting via an increase in IGF-I secretion. Support for this proposal comes from the finding that, in fish, which like the fetus exhibit a low endogenous rate of prolactin release, IGF-I has a marked stimulatory effect on prolactin secretion.

The extent to which prolactin may be acting directly or via placental lactogen receptors remains an area of debate. It should be noted, however, that immunoneutralisation against placental lactogen has no detectable effects on maternal or fetal metabolism. Moreover, in vivo effects on bone metabolism of 50 ng/ml circulating prolactin have been demonstrated in developing rats in the face of lg/ml quantities of placental lactogen. It is not only prolactin that can have biological effects, as phosphorylation of prolactin has been shown to be physiologically important at different stages of the reproductive cycle. When prolactin promotes cell proliferation, phosphorylated prolactin antagonises this effect.

In contrast, when the function of prolactin is not proliferative, phosphorylated prolactin may be an equal or greater receptor agonist than prolactin. Studies in pregnant rats have demonstrated profound and differential effects of maternal administration with prolactin or phosphorylated prolactin on a large range of fetal tissues, confirming that it is not the absolute level of prolactin that is important, but the ratio of unmodified to phosphorylated prolactin. For example, fetal thymus size was decreased by maternal prolactin administration, but increased by phosphorylated prolactin. The extent to which these effects may contribute to clinically significant effects in response to low plasma prolactin concentrations, including pre-maturity, remains to be explored.

Prolactin and Fetal Energy Balance

Prolactin has also been suggested to regulate energy balance. The demonstration of high expression of uncoupling protein 2 (UCP2) throughout the non-human primate adult brain, including the hypothalamus and pituitary, could be linked to prolactin secretion. In the anterior pituitary, cell expression of UCP2 is widespread, although it is not co-localised with GH-producing cells. It is therefore hypothesised that UCP2 is co-expressed by prolactin-secreting cells, thereby providing a direct link between prolactin and brain metabolism. In the late-gestation fetus, a central indicator of nutrient supply is the abundance of neuropeptide Y.

In the adult, neuropeptide Y is a potent appetite stimulant. Interestingly, the peak in prolactin mRNA abundance in the fetal anterior pituitary precedes the peak in hypothalamic neuropeptide Y. Maternal nutrient restriction has been shown to result in a precocious peak in the fetal plasma prolactin concentration. This may be a critical stimulus for the higher levels of fetal hypothalamic neuropeptide Y mRNA observed following maternal under nutrition, thereby leading to parturition.

UCPs and Energy Balance

Prolactin, leptin and GH have been shown to have direct or indirect effects on fetal and neo-natal energy balance. These include effects on intermediary metabolism and on UCP abundance. Not only the amount of UCP1, but also that of voltage-dependent anion channel (VDAC), peaks at birth. Importantly our antibody does not appear to be cross-reacting with UCP1, and the loss of VDAC in parallel with UCP1 may be indicative of pronounced apoptosis within brown adipocytes. We are unable to detect UCP1 in adipose tissue at 30 days of post-natal age, whereas VDAC is clearly detectable.

The role of VDAC in the both the young and adult mammal is the subject of speculation. It should be further noted that lambs, sheep and humans are all able to maintain normothermia when cold-adapted. This is despite the absence of UCP1 and without visible signs of shivering, which contrasts with the situation in mice, in which the absence of UCP1 or its low abundance is associated with hypothermia. Surprisingly, knockout mice in which UCP1 is completely absent do not become obese, whereas transgenic mice with low UCP1 abundance do become obese.

The extent to which the hormonal regulation of each UCP differs remains to be established. Interestingly, VDAC is detectable in adipose tissue at 1 month of age, when UCP1 is no longer present. This coincides with the large increase in fat mass, and may represent a transition in mitochondrial function from heat generation to regulation of the production of reactive oxygen species. Rapid loss of UCPs in a variety of tissues may have a beneficial role, by enabling the post-natal mammal to cope effectively with infection by allowing effective production of reactive oxygen species.

The role of UCP3, which is present in both adipose tissue and skeletal muscle, also remains to be established; overexpression of UCP3 in skeletal muscle is associated with growth retardation, despite an increase in food intake. The extent to which this is a direct response to excess UCP3 abundance in skeletal muscle, as opposed to alterations in fetal development, possibly associated with impaired mitochondrial development, has not been addressed.

8

Assimilation of Food (Animals)

Nutrition deals with the composition of food, its energy content, and slowly (or not at all) synthesised organic molecules. Nutrition, study of the materials that nourish an organism and of the manner in which the separate components are used for maintenance, repair, growth, and reproduction. Nutrition is achieved in various ways by different forms of life. Plants that contain the green pigment chlorophyll can synthesise their food from inorganic substances in the process called photosynthesis.

Organisms such as plants that can thus manufacture complex organic compounds from simple inorganic nutrients are termed autotrophic. Organisms that must obtain "prefabricated" organic compounds from their environment are heterotrophic, and these include the fungi, some other plants, and animals. Heterotrophic plants may be saprophytic or parasitic. Heterotrophic animals may be parasites, herbivores, carnivores (meat eaters), or omnivores (obtaining nutrition from both plants and animals).

A heterotroph (Greek heteron = (an)other and trophe = nutrition) is an organism that requires organic substrates to get its carbon for growth and development. Contrast with autotrophs which use carbon dioxide as sole carbon source. All animals are heterotrophic, as well as fungi and many bacteria. Some parasitic plants have also turned fully or partially heterotrophic, whereas carnivorous plants use their flesh diet to augment their nitrogen supply, but are still autotrophic. Eterotrophs are unable to synthesise organic, carbon based compounds independently from the inorganic environment's sources (e.g. Animalia, unlike Plantae, cannot photosynthesise) and therefore must obtain their nutrition from another heterotroph or an autotroph or a species to be termed a heterotroph, it must obtain its carbon from organic compounds.

If it obtains nitrogen from organic compounds, but not carbon, it will be deemed an autotroph. If a species obtains carbon from organic compounds and a) obtains energy from light or b) obtains energy from the oxidation of inorganic compounds, it is called a photoheterotroph and a chemoheterotroph respectively.

Human Nutrition

Humans require food substances to supply the components necessary to build tissues, to repair tissues as they wear out and die, to keep the body in good working condition, and to supply fuel for energy. For good nutrition a person should eat a well-balanced diet, that is, one that provides an adequate amount of each of the classes of nutrients each day, furnishing at the same time an adequate but not excessive number of calories for the body's energy needs.

Children require relatively larger amounts of nutrients and calories because of their rapid growth. The foods required for proper nutrition fall roughly into three major groups: proteins, carbohydrates, and fats; vitamins, minerals, and water are also important. Chemotrophs are organisms (mostly bacteria) deriving their energy from inorganic chemical reactions. Phototrophs convert sunlight energy into sugar or other organic molecules. Heterotrophs eat to obtain energy from the breakdown of organic molecules in their food.

Macronutrients are foods required on a large scale each day. These include carbohydrates, lipids, and amino acids. Water is essential, correct water balance is a must for proper functioning of the body. About 60 per cent of the diet should be carbohydrates, such as are in milk, meat, vegetables, grains and grain products. The diet should contain at least 100 grams of carbohydrate every day.

Proteins are polymers composed of amino acids. Proteins are found in meat, milk, poultry, fish, cereal grains and beans. They are needed for cellular growth and repair. Twenty amino acids are found in proteins, of which humans can make eleven. The remaining nine are the essential amino acids which must be supplied in the diet.

Normally proteins are not used for energy, however during starvation muscle proteins are broken down for energy. Excess protein can be used for energy or converted to fats. Lipids and fats generate the greatest energy yield, so many plants and animals store energy as fats. Lipids and fats are present in oils, meats, butter, and plants (such as avocado and peanuts). Some fatty acids, such as linoleic acid, are essential and must be included in the diet. When present in the intestine, lipids promote the uptake of vitamins A, D, E, and K.

Vitamins are organic molecules required for metabolic reactions. They usually cannot be made by the body and are needed in trace amounts. Vitamins may act as enzyme co-factors or co-enzymes. Some vitamins are soluble in fats, some in water. Minerals are trace elements required for normal metabolism, as components of cells and tissues, and for nerve conduction and muscle contraction. They can only be obtained from the diet. Iron (for haemoglobin), iodine (for thyroxin), calcium (for bones), and sodium (nerve message transmission) are examples of minerals. There is a quantitative relationship between nutrients and health. Imbalances can cause disease. Many studies have concluded nutrition is a major factor in cardiovascular disease, hypertension, and cancer.

Human Dietary

The human diet must provide the following:

- *Calories;* enough to meet our daily energy needs.
- *Smino acids.* There are nine, or so, "essential" amino acids that we need for protein synthesis and that we cannot synthesise from other precursors.
- *Fatty acids.* There are three "essential" fatty acids that we cannot synthesise from other precursors.
- *Minerals.* Inorganic ions. We probably need 18 different ones: a few like calcium in relatively large amounts; most, like zinc, in "trace" amounts.
- *Vitamins. A* dozen, or so, small organic molecules that we cannot synthesise from other precursors in our diet. Inadequate intake of some vitamins produces easily-recognised deficiency diseases like
  — Scurvy: lack of ascorbic acid (vitamin C)
  — Beriberi: lack of thiamine (vitamin B1)
  — Pellagra: lack of niacin.

However, it is so difficult to exclude some other possible vitamins from the diet that deficiency diseases are hard to demonstrate. Some minerals are needed in such vanishingly small amounts that it is practically impossible to prepare a diet that does not include them. However, totally synthetic diets are now available for intravenous feeding of people who cannot eat. This so-called total parenteral nutrition has revealed, unexpectedly, some additional trace element needs: chromium and molybdenum.

Carbohydrates : Carbohydrates provide a readily available energy source. Surplus carbohydrates are also converted by the body to glycogen and fat, the storage forms of calories for energy, and to some of the amino acids used in protein synthesis. Most health professionals recommend that carbohydrates comprise 50 per cent to 60 per cent of the dietary calories, of which most (c.80 per cent of all carbohydrates eaten) should be complex carbohydrates, such as cereals and vegetables.

Complex carbohydrates are preferred because the fast-acting simple carbohydrates, such as honey and sugar, are difficult for the body (especially the pancreas) to handle in large doses. Simple carbohydrates also lack the vitamins, minerals, proteins, and fibre that generally accompany foods rich in complex carbohydrates. Cereals, fruits, vegetables, legumes, and pasta are good sources of complex carbohydrates. In some poor countries, too many children do not receive enough calories to grow properly. In order to maintain blood sugar levels, they attack their own protein. This condition of semi-starvation is known as marasmus.

Protein in the Diet : Protein in the diet provides amino acids for forming body proteins, including the structural proteins for building and repairing tissues, and the enzymes for carrying out the metabolic processes. In addition, protein may be used as a source of energy when the preferred fat and carbohydrate supply runs low. A body that is in the process of building itself (such as that of a growing child or an adult recovering from illness) will need a greater proportion of protein to weight than one that is fully grown and utilises protein merely for repair of worn-out tissues.

The average adult requires 1 gram of protein per kilogram of body weight per day; children may require two to three times this amount. Human proteins consist of up to 22 different amino acids, of which 9 (called essential amino acids) must be supplied by food protein; the other 13 are synthesised by human cells. Complete protein sources-those foods containing all 22 amino acids-include animal products such as meat, eggs, cheese, and milk.

Incomplete protein sources, such as vegetables, beans, and

grains, may be combined to create complete proteins. Humans must include adequate amounts of 9 amino acids in their diet. These "essential" amino acids cannot be synthesised from other precursors. However, cysteine can partially meet the need for methionine (they both contain sulphur), and tyrosine can partially substitute for phenylalanine.

Essential Amino Acids
Histidine
Isoleucine
Leucine
Lysine
Methionine (and/or cysteine)
Phenylalanine (and/or tyrosine)
Threonine
Tryptophan
Valine

Two of the essential amino acids, lysine and tryptophan, are poorly represented in most plant proteins. Thus strict vegetarians should take special pains to ensure that their diet contains sufficient amounts of these two amino acids. Birds, mammals, and some other animals are able to discriminate food that contains a nutrient, e.g., an essential amino acid, that they need from food that doesn't. If offered a food lacking that nutrient, they quickly stop eating it.

In rats, at least, it turns out that certain neurons in the brain detect the lack of an essential amino acid and signal the appetite centres of the brain to stop feeding on deficient food. The neurons detect the lack by the failure of their transfer RNAs (rRNAs) for that amino acid to acquire it. Rats whose tRNAs for threonine have been blocked from loading threonine cease feeding even if their food contains adequate concentrations of it.

Fats : Fats in the diet provide a concentrated source of energy; 1 gram of fat supplies about 9 calories as opposed to only 4 calories per gram of carbohydrates and protein. Fats in the body, in addition to acting as a source of stored energy, supply physical protection and insulation for tissues and form important portions of cell membrane structure.

Fats also aid in the absorption of the fat-soluble vitamins (vitamins A, D, E, and K) from the intestine. Milk, butter, meat, and oils are important sources of fat.

Ingested fats provide the precursors from which we synthesise our own fat as well as cholesterol and various phospholipids. Fat provides our most concentrated form of energy. Its energy content (9 kcal/gram) is over twice as great as carbohydrates and proteins (4 kcal/gram).

Humans can synthesise fat from carbohydrates (as most of us know all too well!). However, three essential fatty acids cannot be synthesised this way and must be incorporated in the diet. These are:

- Oleic acid,
- Linolenic acid,
- Arachidonic acid.

All are unsaturated; that is, have double bonds.

Types of fats

— *Saturated:* No double bonds between the carbon atoms in the fatty acid chains. Most animal fats (e.g., butter) are highly saturated.

— *Monounsaturated:* Have a single double bond in the fatty acid chains. Examples are olive, peanut, and rapeseed (canola) oil.

— *Polyunsaturated:* Have two or more double bonds in their fatty acid chains. Examples: corn, soy bean, cottonseed, sunflower, and safflower oils.

— *Trans Fats:* Have been partially hydrogenated producing

— Fewer double bonds and, of those that remain,

— . Converting them from a cis to a trans configuration.

- *Omega-3 fats:* Have at least one double bond three carbon atoms in from the end of the fatty acid molecule. Linolenic acid is an example. Fish oils are a rich source of omega-3 fatty acids.

The relationship between fat in the diet and cardiovascular disease, there is still no consensus, but the evidence seems to indicate that:

- A diet high in fat is harmful.
- Mono and polyunsaturated fats are less harmful than saturated ones, except that
- trans unsaturated fats may be worse than saturated fats.
- Ingestion of omega-3 unsaturated fats may be protective.

At present, food labels in the U.S. list the total amount of fat in a serving of the product with a breakdown of the amounts of saturated (1 g), polyunsaturated (0.5 g), and monounsaturated fat (1.5 g). There is a proposal to have them included, but at present they are not. However, if you add the amounts of saturated, polyunsaturated, and monounsaturated fat, and the total does not equal "Total Fat", the discrepancy (2 g in this example) represents the amount of trans fat. Baked goods tend to have quite a bit of trans fat.

Vitamins and Minerals : To keep the body functioning properly it is necessary to have, in addition to the basic foods, a sufficient intake of accessory substances such as vitamins, minerals, and enough water to carry nutrients to the tissues and waste products away from them. A minimum of about 2 litres of liquid per day are recommended for the average adult. Vitamins function as co-enzymes in important body processes, with the exception of vitamin D, which is synthesised upon exposure to sunlight.

A large variety of minerals are required, some in trace amounts and others, such as calcium and iron, in relatively large amounts. Milk, cheese, and dark, leafy green vegetables are excellent sources of calcium; liver, meat, and egg yolks are good sources of iron. Minerals are vital to the development of teeth and bones (calcium, phosphorus, and fluoride) and to the functioning of a number of the body's metabolic systems.

Iron is a necessary part of haemoglobin in the blood; various metals are required in many enzymes; sodium and potassium are essential to maintenance of fluid balance and functioning of the nervous system; magnesium is needed for the normal functioning of nerves and muscles; and iodine is required for thyroid hormone. The usefulness of vitamin and mineral supplements for a person of good health who eats a well-balanced diet continues to provoke debate among health experts. Types of minerals are discussed below:

Calcium : Calcium is essential to almost every function in the body. For most of these, such as:

- Blood clotting,
- Intracellular signaling,
- Muscle contraction,

Only trace amounts are needed. However, large amounts of calcium are needed to make bone (which is 18 per cent calcium), So substantial amounts are needed in the diet, especially during infancy, childhood, and pregnancy. Three hormones:

- Parathyroid hormone (PTH)
- Calcitonin, and
- Calciferol (vitamin D)

Work together to regulate how much calcium

- is absorbed from your food
- is taken from or added to bone
- is excreted in the urine.

A temporary deficit in the amount of calcium in the diet can be compensated for by its removal from the huge reserves in bone.

Iron (Fe) : Iron is incorporated in a number of body constituents, notably

- Cytochromes
- Myoglobin, and
- Hemoglobin.

Not surprisingly, an iron deficiency shows up first as anaemia. In developed countries like the U.S., iron deficiency is the most common mineral deficiency. It is particularly common among women because of

- the loss of blood during menstruation
- the need for extra iron during pregnancy and breast feeding.

Marginal iron intake is so widespread that some nutritionists want to have iron added to common foods like bread and cereals, just as some vitamins now are. However, excess iron in the body also leads to problems, and this has made the proposal controversial. Even iron supplement tablets pose risks: thousands of children in the U.S. are accidentally poisoned each year by swallowing too many iron tablets. In fact, iron is the most frequent cause of poisoning deaths among children in the U.S.

Iodine

- Incorporated in the hormones thyroxine (T4) and triiodothyronine (T3).
- In regions with iodine-deficient soils, food may not contain enough iodine to meet body needs. The result is goiter: a swelling of the thyroid gland.
- The use of iodised salt (table salt to which a small amount of sodium iodide, KI, is added) has reduced the incidence of goiter in most developed countries.

Fluoride : The value of fluoride (in ionised form, F-) was first recognised as a preventive for dental caries (cavities). This makes sense because fluoride ions are incorporated along with calcium and phosphate ions in the crystalline structure of which both bones and teeth are constructed. But it may have other functions. In order to grow properly, a rat must consume 0.5 parts per million (ppm) of fluoride ions in its diet.

Humans get most of their fluoride in drinking water. In regions where the natural amount is less than 1 ppm, many communities add enough fluoride to bring the concentration up to 1 ppm.

Perhaps because the range between optimum and excess is more narrow for fluoride than for most minerals in the diet, water fluoridation has been controversial. Leaving aside the philosophical and political questions raised by proponents and opponents of fluoridation, the safety and efficacy of this public health measure has been thoroughly established.

Zinc

Zinc is incorporated in many:

- Enzymes and
- Transcription factors

Zinc supplements are popular for their supposed antioxidant properties and to hasten the recovery from colds. Excessive intake of zinc causes a brief illness. Its most frequent cause is from ingested acidic food or drink that has been stored in galvanised (zinc-coated) containers.

Different kinds of vitamins are discussed below:

Vitamin A (Retinol)

- Functions: Multiple, including serving as the precursor to retinal, the prosthetic group of all four of the light-absorbing pigments in the eye and regulating gene expression essential for the health of epithelia.
- Sources: cream, butter, fish liver oils, eggs. Carrots and

some other vegetables provide beta-carotene, which the liver can convert into vitamin A.

- Deficiency: night-blindness.
- Excess: stored in the liver, but can be toxic in large doses, especially in children. Even in adults the range between too little and too much is narrow: ingesting vitamin A in amounts not much greater than the recommended dietary allowance (RDA) leads to an increase in bone fractures later in life. High doses taken early in pregnancy have been linked to a greater risk of birth defects.

Vitamin B_1 (Thiamine)

- Function: coenzyme in cellular respiration.
- Sources: meat, yeast, unpolished cereal grains, enriched bread and breakfast cereals.
- Deficiency: beriberi. Rarely found in developed countries except among alcoholics.
- Excess: water soluble and any excess easily excreted.

Vitamin B_2 (Riboflavin)

- Function: prosthetic group of flavouprotein enzymes used in cellular respiration.
- Sources: liver, eggs, cheese, milk, enriched bread and breakfast cereals.
- Deficiency: damage to eyes, mouth, and genitals.
- Excess: water soluble and any excess easily excreted.

Vitamin B_3 (Nicotinic acid or Niacin)

- Function: this member of the B vitamins is a precursor of NAD and NADP
- Sources: meat, yeast, milk, enriched bread and breakfast cereals.
- Deficiency: pellagra (producing skin lesions); a risk where corn is the staple carbohydrate.

- Excess: accidental ingestion of very high doses produces a brief illness, but niacin is water soluble and any excess is quickly excreted.

Vitamin B_{12}

- Function: needed for DNA synthesis.
- Sources: liver, eggs, milk; needs intrinsic factor to be absorbed.
- Deficiency: pernicious anemia; caused by lack of intrinsic factor or a vegan diet.
- Excess: none identified.

Folk acid (Folacin)

- Function: synthesis of purines and pyrimidines.
- Sources: green leafy vegetables, but destroyed by cooking.
- Deficiency: anemia, birth defects. Women who expect to become pregnant should be extra careful that they receive adequate amounts (400 ug/day).
- Excess: water soluble and any excess easily excreted.

Vitamin C (Ascorbic acid)

- Functions: coenzyme in the synthesis of collagen.
- Sources: citrus fruits, green peppers, tomatoes; destroyed by cooking.
- Deficiency: scurvy.
- Excess: Many people take huge amounts of vitamin C, hoping to ward off colds, cancer, etc. They seem to suffer no harm except, perhaps, to their wallets.
- However, in vitro studies show that a level of vitamin C equivalent to a person consuming 200 mg a day converts some lipids into substances that can damage DNA.

Vitamin D

- Functions: absorption of calcium from the intestine and bone formation.

- Sources:
- synthesised when ultraviolet light (mostly UV-B) strikes the skin (forms vitamin D3).
- Present in fish liver oils, butter, and steroid-containing foods irradiated with ultraviolet light (vitamin D2).
- Deficiency:
  — rickets - inadequate conversion of cartilage to bone- in children;
  — osteomalacia - softening of the bones - in adults.

Until recently, rickets has been very rare in North America. But the combination of two growing trends

- Breast feeding and
- Protecting children from exposure to the sun has caused cases to reappear especially in northern latitudes with their short winter days.

Breast milk provides less than 20 per cent of the recommended daily dose for infants. Until the infant is old enough to eat foods fortified with vitamin D, many paediatricians recommend vitamin supplements for breast-fed babies.

- Excess: However, this fat-soluble vitamin is dangerous in very high doses, especially in infants, causing excessive calcium deposits and mental retardation. So some pediatricians view the use of vitamin D supplements for infants with caution (especially since certain preparations have been found to contain amounts far higher than that listed on the label).

Vitamin E (Tocopherol)

- Function: acts as an antioxidant agent in cells.
- Sources: vegetable oils, nuts, spinach.
- Deficiency: anemia, damage to the retinas.
- Excess: high doses may be toxic.

Vitamin K

- Function: needed for the synthesis of blood clotting factors.
- Sources: spinach and other green leafy vegetables; synthesised by intestinal bacteria.
- Deficiency: slow clotting of blood.
- Excess: high doses may be toxic in infants.

"Natural" versus "Synthetic" Vitamins

- There is no scientific distinction between them.
- The thiamine molecule (or any other molecule) is the same entity
    - whether synthesised by a plant or by an organic chemist
    - whether it is still in plant or animal material or has been extracted and incorporated in a pill.

Role of Good Nutrition

Good nutrition is reflected not only in the growth and function of the body but also in its appearance. The eyes, skin, hair, and teeth indicate whether body nourishment is good or poor. A poorly nourished child will fail to grow properly; a poorly nourished adult will have a decreased resistance to infection and disease. A diet deficient in proteins causes a disease called kwashiorkor in children; a diet deficient in both protein and calories results in marasmus, with lethargy, abdominal enlargement, and wasting-the classical malnutrition syndrome. Poor nutrition may result from excesses in the diet as well as deficiencies; excess of certain vitamins or minerals can produce potentially lethal disease states, and excess of carbohydrates or fat can result in obesity.

Specialised Diets

Specialised diets are useful in the treatment of certain disease states; the most common is a low-calorie diet to produce weight loss in obese persons. A diet low in phenylalanine is used to treat phenylketonuria. A diet low in cholesterol and saturated fats seems to be useful in the treatment and prevention of heart disease. Elimination of certain foods from the diet may be necessary to control allergies in some individuals. In all cases, however, specialised diets must provide all classes of essential nutrients in adequate amounts to maintain health in adults and support growth in children.

Digestive System

Single-celled organisms can directly take in nutrients from their outside environment. Multicellular animals, with most of their cells removed from contact directly with the outside environment, have developed specialised structures for obtaining and breaking down their food.

Animals depend on two processes: feeding and digestion. Animals are heterotrophs, they must absorb nutrients or ingest food sources. Ingestive eaters, the majority of animals, use a mouth to ingest food. Absorptive feeders, such as tapeworms, live in a digestive system of another animal and absorb nutrients from that animal directly through their body wall. Filter feeders, such as oysters and mussels, collect small organisms and particles from the surrounding water. Substrate feeders, such as earthworms and termites, eat the material (dirt or wood) they burrow through. Fluid feeders, such as aphids, pierce the body of a plant or animal and withdraw fluids.

The digestive system uses mechanical and chemical methods to break food down into nutrient molecules that can be absorbed into the blood. There are two types of plans and two locations of digestion. Sac-like plans are found in many invertebrates, who have a single opening for food intake and the discharge of wastes. Vertebrates use the more efficient tube-within-a-tube plan with food entering through one opening (the mouth) and wastes leaving through another (the anus).

- Intracellular digestion: food is taken into cells by phagocytosis with digestive enzymes being secreted into the phagocytic vesicles; occurs in sponges, coelenterates and most protozoans.
- Extracellular digestion: digestion occurs in the lumen (opening) of the digestive system, with the nutrient molecules being transferred to the blood or body fluid; occurs in chordates, annelids, and crustaceans.

Stages in Digestive System

- Movement: propels food through the digestive system
- Secretion: release of digestive juices in response to a specific stimulus
- Digestion: breakdown of food into molecular components small enough to cross the plasma membrane
- Absorption: passage of the molecules into the body's interior and their passage throughout the body
- Elimination: removal of undigested food and wastes

Three phases occur during what we loosely refer to as "digestion". Digestion proper, which is the mechanical and chemical breakdown of food into particles/molecules small enough to pass into the blood stream. Absorption into the blood stream. Assimilation, the passage of the food molecules into body cells.

Human Digestive System : The human digestive system is a coiled, muscular tube (6-9 metres long when fully extended) extending from the mouth to the anus. Several specialised compartments occur along this length: mouth, pharynx, esophagus, stomach, small intestine, large intestine, and anus. Accessory digestive organs are connected to the main system by a series of ducts: salivary glands, parts of the pancreas, and the liver and gall bladder (bilary system).

Mouth and Pharynx : Mechanical breakdown begins in the mouth by chewing (teeth) and actions of the tongue. Chemical breakdown of starch by production of salivary amylase from the salivary glands. This mixture of food and saliva is then pushed into the pharynx and esophagus. The esophagus is a muscular tube whose muscular contractions (peristalsis) propel food to the stomach.

In the mouth, teeth, jaws and the tongue begin the mechanical breakdown of food into smaller particles. Most vertebrates, except birds (who have lost their teeth to a hardened bill), have teeth for tearing, grinding and chewing food. The tongue manipulates food during chewing and swallowing; mammals have tastebuds clustered on their tongues.

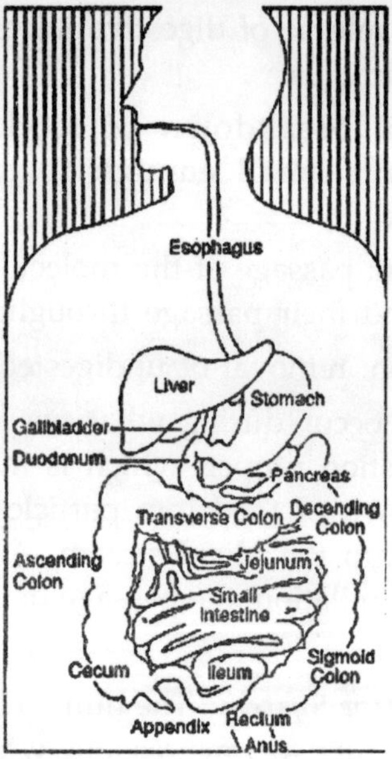

Human Digestive System

Salivary glands secrete salivary amylase, an enzyme that begins the breakdown of starch into glucose. Mucus moistens food and lubricates the esophagus. Bicarbonate ions in saliva neutralise the acids in foods. Swallowing moves food from the mouth through the pharynx into the esophagus and then to the stomach.

- *Step 1 :* A mass of chewed, moistened food, a bolus, is moved to the back of the moth by the tongue. In the pharynx, the bolus triggers an involuntary swallowing reflex that prevents food from entering the lungs, and directs the bolus into the esophagus.
- *Step 2 :* Muscles in the esophagus propel the bolus by waves of involuntary muscular contractions (peristalsis) of smooth muscle lining the esophagus.
- *Step 3 :* The bolus passes through the gastroesophogeal sphincter, into the stomach. Heartburn results from irritation of the esophagus by gastric juices that leak through this sphincter.

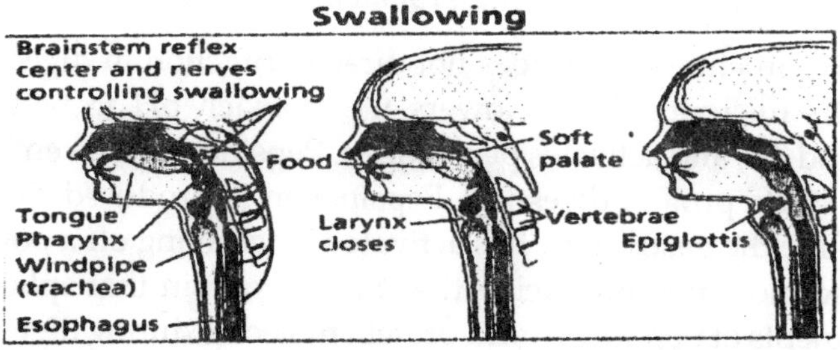

Structure of the throat and the mechanics of swallowing.

Stomach (or Churn, Churn, Churn) : During a meal, the stomach gradually fills to a capacity of 1 litter, from an empty capacity of 50-100 millilitres. At a prize of discomfort, the stomach can distend to hold 2 litres or more.

Epithelial cells line inner surface of the stomach, and secrete about 2 litres of gastric juices per day. Gastric juice contains hydrochloric acid, pepsinogen, and mucus; ingredients important in digestion. Secretions are controlled by nervous (smells, thoughts, and caffeine) and endocrine signals. The stomach secretes hydrochloric acid and pepsin. Hydrochloric acid (HCl) lowers pH of the stomach so pepsin is activated. Pepsin is an enzyme that controls the hydrolysis of proteins into peptides. The stomach also mechanically churns the food. Chyme, the mix of acid and food in the stomach, leaves the stomach and enters the small intestine.

Peristalsis

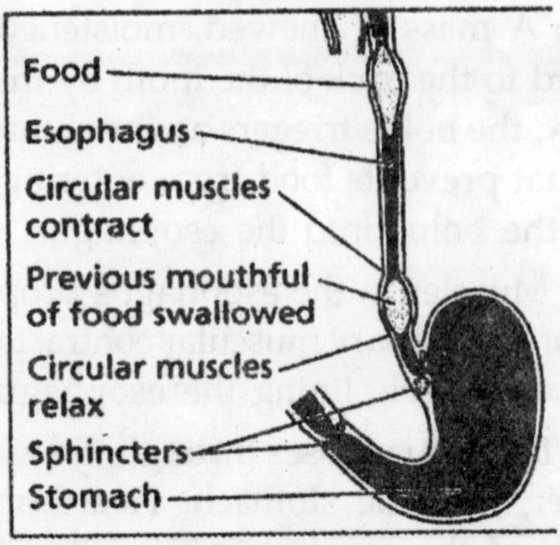

Peristalsis and the movement of food from the mouth to the stomach

Hydrochloric acid does not directly function in digestion: it kills microorganisms, lowers the stomach pH to between 1.5 and 2.5; and activates pepsinogen. Pepsinogen is an enzyme that starts protein digestion. Pepsinogen is produced in cells that line the gastric pits. It is activated by cleaving off a portion of the molecule, producing the enzyme pepsin that splits off fragments of peptides from a protein molecule during digestion in the stomach.

Carbohydrate digestion, begun by salivary amylase in the mouth, continues in the bolus as it passes to the stomach. The bolus is broken down into acid chyme in the lower third of the stomach, allowing the stomach's acidity to inhibit further carbohydrate breakdown. Protein digestion by pepsin begins. Alcohol and aspirin are absorbed through the stomach lining into the blood. Epithelial cells secrete mucus that forms a protective barrier between the cells and the stomach acids. Pepsin is inactivated when it comes into contact with the mucus. Bicarbonate ions reduce acidity near the cells lining the stomach. Tight junctions link the epithelial stomach-lining cells together, further reducing or preventing stomach acids from passing.

Ulcers : Peptic ulcers result when these protective mechanisms fail. Bleeding ulcers result when tissue damage is so severe that bleeding occurs into the stomach. Perforated ulcers are life-threatening situations where a hole has formed in the stomach wall. At least 90 per cent of all peptic ulcers are caused by Helicobacter pylori. Other factors, including stress and aspirin, can also produce ulcers.

Small Intestine : The small intestine is where final digestion and absorption occur. The small intestine is a coiled tube over 3 metres long. Coils and folding plus villi give this 3m tube the surface area of a 500-600m long tube. Final digestion of proteins and carbohydrates must occur, and fats have not yet been digested. Villi have cells that produce intestinal enzymes which complete the digestion of peptides and sugars.

The absorption process also occurs in the small intestine. Food has been broken down into particles small enough to pass into the small intestine. Sugars and amino acids go into the bloodstream via capillaries in each villus. Glycerol and fatty acids go into the lymphatic system. Absorption is an active transport, requiring cellular energy. Food is mixed in the lower part of the stomach by peristatic waves that also propel the acid-chyme mixture against the pyloric sphincter.

Increased contractions of the stomach push the food through the sphincter and into the small intestine as the stomach empties over a 1 to 2 hour period. High fat diets significantly increase this time period.

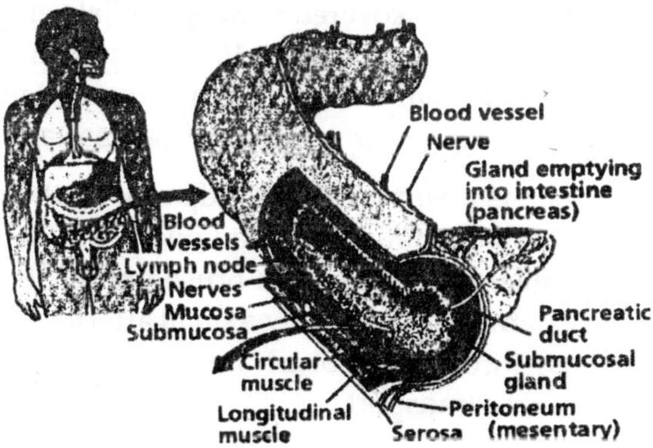

Structure and details of the small intestine

The small intestine is the major site for digestion and absorption of nutrients. The small intestine is up to 6 metres long and is 2-3 centimetres wide. The upper part, the duodenum, is the most active in digestion. Secretions from the liver and pancreas are used for digestion in the duodenum. Epithelial cells of the duodenum secrete a watery mucus. The pancreas secretes digestive enzymes and stomach acid-neutralising bicarbonate. The liver produces bile, which is stored in the gall bladder before entering the bile duct into the duodenum.

Digestion of carbohydrates, proteins, and fats continues in the small intestine. Starch and glycogen are broken down into maltose. Proteases (enzymes secreted from the pancreas) continue the breakdown of protein into small peptide fragments and some amino acids.

Bile emulsifies fats, facilitating their breakdown into progressively smaller fat globules until they can be acted upon by lipases. Bile contains cholesterol, phospholipids, bilirubin, and a mix of salts. Fats are completely digested in the small intestine, unlike carbohydrates and proteins.

Most absorption occurs in the duodenum and jejeunum (second third of the small intestine). The inner surface of the intestine has circular folds that more than triple the surface area for absorption. Villi covered with epithelial cells increase the surface area by another factor of 10. The epithelial cells are lined with microvilli that further increase the surface area; a 6 metre long tube has a surface area of 300 square metres.

Each villus has a surface that is adjacent to the inside of the small intestinal opening covered in microvilli that form on top of an epithelial cell known as a brush border. Each villus has a capillary network supplied by a small arteriole. Absorbed substances pass through the brush border into the capillary, usually by passive transport.

Maltose, sucrose, and lactose are the main carbohydrates present in the small intestine; they are absorbed by the microvilli. Starch is broken down into two-glucose units (maltose) elsewhere. Enzymes in the cells convert these disaccharides into monosaccharides that then leave the cell and enter the capillary. Lactose intolerance results from the genetic lack of the enzyme lactase produced by the intestinal cells. Peptide fragments and amino acids cross the epithelial cell membranes by active transport. Inside the cell they are broken into amino acids that then enter the capillary. Gluten enteropathy is the inability to absorb gluten, a protein found in wheat.

Digested fats are not very soluble. Bile salts surround fats to form micelles that can pass into the epithelial cells. The bile salts return to the lumen to repeat the process. Fat digestion is usually completed by the time the food reaches the ileum (lower third) of the small intestine. Bile salts are in turn absorbed in the ileum and are recycled by the liver and gall bladder. Fats pass from the epithelial cells to the small lymph vessel that also runs through the villus.

Liver and gall bladder : The liver sends bile to the small intestine. Bile contains bile salts, which emulsify fats, making them susceptible to enzymatic breakdown. In addition to digestive functions, the liver functions in other systems:

- Detoxification of blood;
- Synthesis of blood proteins;
- Destruction of old erythrocytes and conversion of hemoglobin into a component of bile;
- Production of bile;
- Storage of glucose as glycogen; and
- Production of urea from amino groups and ammonia.

Glycogen serves as a reservoir for glucose. Low glucose levels in the blood cause release of hormones to stimulate breakdown of glycogen into glucose. When no glucose or glycogen is available, amino acids are converted into glucose in the liver. The process of deamination removes the amino groups from amino acids. Urea is formed and passed through the blood to the kidney for export from the body.

Liver Diseases : Jaundice occurs when the characteristic yellow tint to the skin is caused by excess haemoglobin breakdown products in the blood, a sign that the liver is not properly functioning. Jaundice may occur when liver function has been impaired by obstruction of the bile duct and by damage caused by hepatitis. Hepatitis A, B, and C can all cause liver damage. Cirrhosis of the liver commonly occurs in alcoholics, who place the liver in a stress situation due to the amount of alcohol to be broken down.

Pancreas : The pancreas sends pancreatic juice, which neutralises the chyme, to the small intestine through the pancreatic duct.

Large Intestine : The large intestine is made up by the colon, cecum, appendix, and rectum. Material in the large intestine is mostly indigestible residue and liquid. Movements are due to involuntary contractions that shuffle contents back and forth and propulsive contractions that move material through the large intestine. Secretions in the large intestine are an alkaline mucus that protects epithelial tissues and neutralises acids produced by bacterial metabolism.

Water, salts, and vitamins are absorbed, the remaining contents in the lumen form feces (mostly cellulose, bacteria, bilirubin). Bacteria in the large intestine, such as E. coli, produce vitamins (including vitamin K) that are absorbed.

Appetite Regulation

The hypothalamus in the brain has two centres controlling hunger. One is the appetite centre, the other the satiety centre. Gastrin, secretin, and cholecystokinin are hormones that regulate stages of digestion. Protein in the stomach stimulates secretion of gastrin, which causes increased stomach acid secretion and mobility of the digestive tract to move food.

Food passing into the duodenum causes the production of secretin, which in turn promotes release of alkaline secretions from the pancreas, stops further passage of food into the intestine until the acid is neutralised.

Cholecystokinin (CCK) is released from intestinal epithelium in response to fats, and causes the release of bile from the gall bladder and lipase (a fat digesting enzyme) from the pancreas.

Glucose Metabolism : Glucose levels in the blood remain fairly stable. The liver absorbs glucose from the blood and stores it as the polysaccharide glycogen. Blood glucose levels are maintained between meals by releasing glucose from glycogen back into the blood.

9

Developmental Process (Plants)

Development is not simply unfolding a prescribed programme. There is a lot of communication between parts of the organism, and the environment can also influence the developmental process. As the body is developing, self-organising processes arise from the interactions of cells locally, and of tissues on a larger scale. Probabilistic effects can largely influence the position and differentiation of individual cells and contacts between cells. A problem that should be solved by most organisms is how to channel stochastic, self-organising processes into one, coherent process that would produce a well-functioning individual In addition to internal uncertainties, external effects may also influence development.

Protection of the embryo from the environment is an important task, and is solved in various ways by different species. On the other hand, the environment is not only a source of disturbance to development, but can guide development as well. The end-product of development has to match to the habitat in which the organism lives. More precisely, fitness of a genotype is a function of the phenotype and the environment. Therefore, pieces of information that arrive from the environment can be utilised for increasing fitness. An important keyword at the core of this problem is phenotypic plasticity.

Phenotypic plasticity occurs when the same genotype can be expressed in various phenotypes depending on the environment. Plasticity of development has been demonstrated in a number of species for a large variety of traits. Nevertheless, the selective consequences of plasticity are not straightforward to assess. Beside the trivial advantage of adjusting the state of organism to the state of environment, there are potential costs and limitations to plasticity that diminish its adaptive advantage, or may even select for non-plastic (rigid) development. It is not trivial to ask, to what extent is a particular developmental process controlled internally, by the genetic system, and externally, by the environment.

Modular Development

The plant body is essentially modularly organised. The smallest, elementary unit of construction in the shoot system is called a 'metamer', look at the figure below, [a] internal control, [b] external control by the same individual, [c] external control by another individual (denoted by gray). The dotted region represents an Integrated Physiological Unit. Metamers make branches, and branches construct branching systems in a hierarchical manner. Each shoot apex is a potential point for further development.

Development is an iterative process: the same, finite bit of programme is repeated to add new metamers. This add-on nature of development implies that the developmental programme is essentially open-ended. Empirical studies have suggested that the number of iterations can be infinite in some species. In others (for example, in annual plants) iterations certainly terminate, but the number of iterations is not strictly defined; it can depend on the environment. A trivial consequence of iterative development is that the number of organs is not fixed. For example, the number of leaves of a tree can vary with age and habitat conditions.

The first reason for this capability is that organogenesis (formation of organs) is continuous, not restricted to àn embryonic stage. Secondly, organs or even larger parts of the body (metamers or branches) are disposable: they can die without the death of the whole organism.

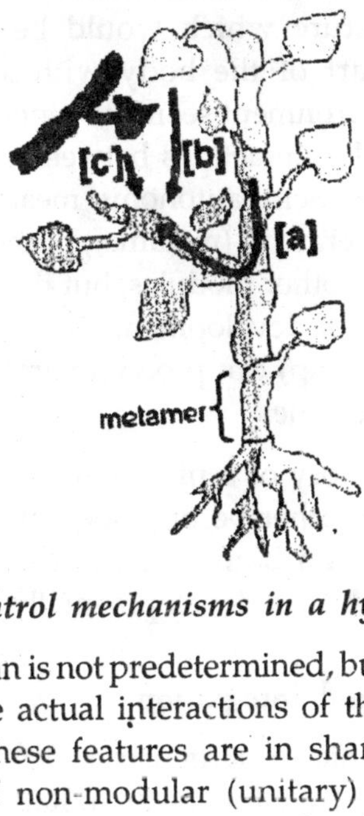

Types of control mechanisms in a hypothetical plant

The body plan is not predetermined, but can flexibly change according to the actual interactions of the organism with its environment. These features are in sharp contrast with the development of non-modular (unitary) organisms, such as vertebrates or arthropods. Iterative development and flexible design make plants particularly suitable for studying phenotypic plasticity. Experiments can use multiple units of the same genotype, and can examine whether these units would express different phenotypes according to external stimuli from the environment.

Modules can be separated and treated individually. Moreover, spontaneous cloning is a frequent phenomenon among plants. In such clonal species, researchers can gain genetically identical, physiologically autonomous individuals (ramets) for the experiments. Thus it can be directly studied how different environments would induce different phenotypes, and how each of these phenotypes is selected locally by its own environment or, being transplanted, by another type of environment.

Plants lack any central controller for morphogenesis or behaviour. There is no central nervous system or any other morphological structure which would be responsible for coordinating one part of the body with another. Indeed, adaptation to the environment seems to hinge on the opposite of centralised control: interactions between individual, semi-autonomous modules. Semi-autonomy means that a module on any level of the hierarchy (metamer, branch, etc.) is partly independent from the other modules, but the independence is not complete. Events of development (such as adding a new metamer to a growing tip) can proceed parallelly, at multiple locations at the same time.

Therefore, life history of a plant is not a single sequence of events, but might rather be described as a set of parallel histories. For example, a branch can develop flowers when another branch is still in a developmentally juvenile stage.

Local Mechanisms and Integration

The domain in which a particular control mechanism can act is an interesting subject for research. Radioactive labelling and other experimental techniques have shown that the pattern of these domains can be fairly sophisticated. A primary determinant is the pattern of vascular contacts, but states of the modules also matter. State transitions and environmental influences can widen up or narrow down the range of integration.

For example, when a branch gets defoliated (e.g. due to herbivore attack), its previous physiological connections to other branches may get re-activated, thus, the Integrated Physiological Units (IPU) becomes wider. On the other hand, there are species in which similar treatments cannot change the borderline between IPUs, i.e., sectoriality is strict. An important conclusion from the experiments is that IPUs do not necessarily coincide with developmental modules. Some modules can be strongly connected physiologically, while others may be independent. This holds for any level of the hierarchy from metamers up to systems of branches.

The highest degree of module independence is achieved by clonal plants, in which each module can develop a self-supporting shoot and root system, and thus, the genetic individual (genet) can get fragmented into independent physiological individuals (ramets). It is remarkable that ramet formation can happen at any level of the developmental hierarchy from metamers (as in Trifolium repens) through branches up to systems of branches. Let us take the example of a branch that has just started its growth, and thus, consists of a single, undeveloped metamer. This initial branch is certainly not capable for self-maintenance, and relies on carbon import from at least one other branch.

Later when the carbon budget becomes positive, the branch can become autonomous. For example, in Capsicum annuum, each adult branch has been found to be selfsupporting: maturing fruits received assimilate only from leaves on the same branch, and not through the main stem. A considerable degree of autonomy can develop even on the level of single metamers. For example, seeds ripening on Pisum arvense stems have been found to receive approximately 66 per cent of their carbon supply from the leaves and pod within their own metamer. Let us consider the import of essential resources other than carbon.

Water and mineral nutrients have to be taken up from the soil. Basically, there are two options. The branch in these hypothetical example would either rely on the existing vascular pathway of resource transport from the root through other branches, or can develop its own root system by forming adventitious roots. The second can lead to full autonomy for all resources, i.e., the case of clonal growth. Experiments suggest that shoot-root connections can also be sectorial. In that case, the branch would not have access to resources from the whole root system, only from a part of it.

Means of Control

Development can be controlled by two types of substances: resources and hormones. Resources (carbon, mineral nutrients and water) act directly by limiting growth. Modules are competing for these resources. Those modules in which consumption is relatively faster create a concentration gradient locally. This gradient is a potential driving force for the transport. But this potential is not necessarily manifested.

The pattern of vascular connections constrains interactions between modules, and it is the relative states of the modules that finally determines the direction and magnitude of transport. Hormones can modify the pattern of resource transport on a longer time scale. For example, auxin production by actively growing shoots determines the direction of vascular differentiation. Besides affecting the connections between modules, hormones can also influence state transitions in each module. For example, gibberellins have been suggested initiate a transition from vegetative to reproductive shoot type in Hedera helix.

Auxin seems to play a central role in the transfer of information about one part of the plant to another part. The most remarkable example is apical dominance. A developing shoot apex releases auxin, which flows basipetally, and inhibits lateral branching (i.e. activation of lateral buds) in the neighbouring metamers. The regulatory process is sophisticated: distance from the roots (due to the action of cytokinins) and age of a bud also matters. Any decline in the quality of the apex, for instance, due to shading, would lead to a decrease in auxin production, and thus, to release of the lateral buds from suppression.

Thus, further growth in the shaded area can be slowed down or even completely abandoned, and a new, lateral direction of growth can be tried. This behaviour can be adaptively advantageous in a wide range of patchy habitat types. An important difference between resources and hormones is that resources arrive from the external environment, and thus, once taken up, can only be depleted. In contrast, hormones are synthesised internally, so, there is practically no theoretical limit on their availability.

Hormonal actions are finer-tuned than the effects of resources: their concentrations are smaller by magnitudes, and even a small change in hormone concentration may cause large change in the phenotype. The effects can be targeted, and can be specific to organs, tissues, or even cell types. In addition, hormones can act before a change in the resource supply would occur, thus, the plant can utilise anticipatory signals. For example, a leaf primordium produces auxin, which causes vascular elements differentiate into the direction of the primordium, where sufficient resource supply would be needed in the future, when the leaf starts to grow.

As a further example for the adaptive significance of anticipation, Novoplansky et al. demonstrated that young Portulaca oleracea plants can sense neighbouring plants from the change of light quality (R/FR ratio), and can respond to this signal by growing away from the direction of the neighbour. Moreover, the response can occur before any significant shading, i.e., before any decline in the amount of resource. Thus, the change in R/FR is a specific, anticipatory signal, and the hormonal system can transmit this to a morphogenetic process. Another example for a predictive signal is photoperiod, which can be used for initiating anthesis before the season for pollination would arrive. This is advantageous, because developing a flower takes time. The distribution of resources (source-sink relations between modules) would provide a sufficient explanation for plant development. In contrast, Sachs lists three basic phenomena which support the importance of hormones.

(1) Early development in juvenile individuals can proceed in a well-balanced way (maintaining a proper shoot:root ratio) even in the absence of some relevant resources: light or nutrients. Water is necessary, but it does not have to be applied through the roots.
(2) Treatments by hormones have been demonstrated to show significant effects on morphogenesis, e.g., on the differentiation of vascular channels.
(3) There are cases when a plant part is a sink for resources but a source for hormones.

The most extreme example is a bacterium-induced tumour, which consumes substrates from other parts of the plants, and is known to release hormones. Local, exogenous application of auxin and cytokinin in a proper ratio could produce the same, over-dominant growth and teratological development. Another, less extreme example for point (3) is apical dominance that is ubiquitous in plants.

An interesting, clearly hormonally controlled process is a release from apical dominance by aR/FR signal. Plants can even be 'cheated' by applying a small quantity of light when it has the proper quality. For example, Deregibus et al. placed light-emitting diodes into closed grassland swards. The diodes emitted a small amount of red light, increasing the R/FR ratio. As a consequence, tussocks in the closed grassland started to branch increasingly, as if the plants were growing in an open area, without any competitor. This response was based on a specific mechanism for signal perception and transaction, that is linked to the hormonal system, and not on the small amount of additional light.

Researchers have demonstrated that even a decrease in the amount of light can induce a high-light growth form when the decrease occurs specifically in the FR spectral range, increasing the R/FR ratio. These results underline the importance of hormones in the control of plant development. Control by resources certainly does not suffice when formation of the optimal branching structure implies resource movement against concentration gradients, i.e. when resource-rich branches should get richer.

There are both positive and negative feedback loops in the control of plant development. A simple example for a positive feedback loop is the above-mentioned effect of auxin in a branch. If a branch grows faster, then it releases more auxin. Auxin suppresses the development of other branches, thus, this particular branch would gain competitive advantage. Getting more resources would result in faster growth, and thus, the loop gets closed.

Negative feedback can be exemplified by the correlative growth of root and shoot systems. It has been demonstrated by numerous studies that the growing plant tissues require a fairly stable ratio of carbon:nitrogen supply. Therefore, biomass of the assimilating shoot (that is a source of carbon) and of the root (that is a source of nitrogen) has to be well balanced. The plant solves this task by a negative feedback. The shoot releases auxin, and presumably other signal molecules, which enhance root initiation. At the same time, the root produces cytokinins, which stimulate the development of shoots. Thus we have two negative feedback loops linked: each system (shoot and root) inhibits its own development by giving advantage to its competitor.

Naturally, negative feedback tends to equalise initial differences between organs to make the system converge to a stable state (e.g. optimal shoot:root ratio), while positive feedback leads to runaway growth. A typical example for the second is offered by split-shoot experiments. Seminal shoot is removed from a young (4 days old) Pisum sativum seedling to stimulate growth of the two buds that are in the axils of the cotyledon. The two buds are nearly equal, but there may be a small difference in size by chance. This initial difference is enhanced during development, and growth gets concentrated in one of the branches.

The difference between branches increases with time, and

in most plants, the subordinate branch stops growing within ten days. It is very interesting to experiment with this system to explore how external stimuli (shading, physical damage, etc.) might change the dominance relations between branches. A natural role of this positive feedback is to allocate limited resources into the growth of the stronger branch. More vigorous growth (and thus, more release of auxin) usually indicates better conditions in the local environment. Thus, the plant can preferentially grow towards better sites. The degree of dominance of the stronger branch is a trait that is open to evolutionary optimisation: too strong dominance can be disadvantageous in those environments in which the good (high-light) patches are small or ephemerous, because being in a good site is not a guarantee for staying in a good site, and a too strong positive feedback may make reversals costly or impossible. The optimal degree of dominance in various environments would certainly need more attention from modellers.

Time lags : Manifestation of a phenotypic change may require time. Physiological changes are usually faster than morphological ones. Time delay can be especially important when its scale is similar to or coarser than the scale of environmental fluctuations. In this case, state of the environment at the time of induction may considerably differ from the state of the environment at the time of selection. Thus, the mutual information between the inducing and selecting environment maybe too low for promoting the evolution of plasticity at all.

The degree of inducibility was measured by the probability that an environmental change would cause a corresponding phenotypic change. Spatial movement in a patchy environment can have the same effect as temporal fluctuation. An extreme example for a long time lag is organ pre-formation in buds. In some species, for example in Podophyllum peltatum, a shoot can be induced to develop reproductive versus vegetative buds in year *t*. So, the presence or absence of inflorescence is determined in year t. But it is only year *t*+1 when the bud is used by the plant to grow.

Year *t* +1 can differ from year t in the availability of pollinators, abundance of resources, and many other factors that influence the optimal number of reproductive modules. Nevertheless, the decision cannot be reversed. An additional problem is called 'ontogenetic contingency'. The same environmental signal that arrives at a shoot of Podophyllum peltatum can have a different effect depending on the state of that particular shoot. State of a shoot at time t is a result of past, historical events (at *t* -1, etc.). Therefore, developmental history is an important determinant in any actual developmental move. Of course, if there is no direct transfer of information from the past (e.g. *t* - 2 influencing t directly), than this can still be approached by a first-order Markovian model, regarding the state of the plant at time t as a cumulative result of its past history.

External Control versus Internal Control

Some researchers have suggested to group together all the influences that may affect the development of a particular organ, tissue, or cell as 'environmental' influence. They suggest to differentiate between an 'external environment' (i.e. everything outside of the plant individual) and an 'internal environment' (i.e. everything within the plant individual but outside of the organ, tissue or cell that is in the actual focus). Givnish lists the following examples for 'internal environment': "positional information based on hormone levels or cellular contact, or the indirect effects of other cells or tissues on abiotic factors such as oxygen tension or red/far red (R:FR) ratio".

Let us focus on a single metamer. This metamer can export and import resources and hormones from and to other parts of the plant (arrow [a]). This represents the internal control of its development. At the same time, the metamer can be shaded by other branches of the same plant [b] or of other plant individuals [c]. Shading also controls its development, but [b] and [c] are certainly external. They work by other mechanisms than [a]. In the case of [a], the carrier of information is a resource or a hormone. At [b] and [c], the carrier of information is light. It is interesting to ask to what extent a plant might differentiate between [b] and [c].

Different species have different composition of Photosynthetic pigments. So the species of the shading branch would be likely to matter. But individual distinctions within the same species do not have any empirical support. In case of root systems, there have been interesting indications for self/non-self recognition in some experiments, but the mechanism is unknown, and certainly more evidence is needed.

For control [a], plants can have heritable genetic variation, so, [a] can be subject to evolution. Controls [b] and [c] cannot be influenced by evolutionary processes directly. [c] is completely inaccessible to the genome. [b] is accessible, but only in an indirect way: the genome codes information about the architecture of the plant, and architecture influences the probability that a branch would shade another branch. So, it is not the mechanism of control but the result of control that can be under selection.

The notion of 'internal environment' is very important, because underlines the dynamic and responsive nature of plant development, including the plant's responses to its own state. It is obvious that all the somatic cells of the plant individual share the same genome (disregarding somatic mutations). Therefore, the reaction norm that determines a response is coded in all cells, tissues, organs, etc., although it is not necessarily manifested in all parts of the plant. Fitness of a genotype (a type of reaction norm) depends on the performance of the whole plant individual in its environment.

Therefore, the evolutionary unit is the genetic individual (genet). Survival and reproduction of a single branch is relevant only to the extent at which it increases or decreases fitness of the whole plant individual. Therefore, in an evolutionary sense, it is only the 'external environment' that can be considered as environment.

10

Endocrine Glands

Most glands of the body deliver their secretions by means of ducts. These are the exorcrine glands. Other glands manufacture chemical substances which they secrete into the blood stream for transmission to various "target" tissues. These are the endocrine (Gr., *endon* = within ; *krinein* = to secrete) or ductless glands. The secretions which they produce *do not leave* the gland through a duct or a canal but pass *directly* from the cells into the blood streams. Study of these glands and their hormones is called Endocrinology.

The secretions produced by the endocrine glands are called Hormones. The term *Hormone* was introduced by William Bayliss and Earnest Starling in 1904, which is derived from the Greek *hormon,* meaning to excite, to arouse or to set in motion. These secretions are prompted by a stimulus and have an activating influence on the neighbouring or distant tissues. They constitute a mode of communication between cells, a mode of regulation of cell machinery and a channel for integration and coordination in multicellular organisms.

Enzymes vs. Hormones — Hormones act as body catalyst

resembling enzymes since they are required only in very small amounts and are not used during their catalytic action. However, they differ from enzymes in the following ways:

(1) They are produced in an organ other than that in which they ultimately perform their action.

(2) They are secreted into the blood in response to body requirements.

(3) Structurally, they are not always proteins. They include proteins with molecular weights of 30.000 or less, small polypeptides, single amino acids an steroids.

The principal role of hormones in the body is to maintain constant, within a narrow range, the characters of the environment of the cells in our body. The great French biologist Claude Bernard (1885) called it constancy of the *milieu interior* (internal environment): the English physiologist, Walter Cannon (1920) described it as *homoeostasis*. Our body cells can live and function normally only if the fluid that surrounds them and constitutes their environment' remains more or less constant both in physical character and chemical composition. This fluid is in a constant dynamic flux: substances are being added and removed, physical characteristic like alkalinity or temperature are being altered continually. These are the hormones which are responsible for homeostasis in the wake of such changing situations.

The internal environment of the body is controlled partly by the autonomic nervous system and partly by the endocrine glands. The hormones secreted by the ductless glands are mainly excitatory in nature stimulating activity in other organs.

The endocrine system consists of the following glands:

1 Pituitary gland

1 Thyroid gland

4 Parathyroid glands

2 Adrenals or Supra-renal glands

The islets of langerhans in the pancreas

2 Testes in the male

2 ovaries in the female

Placenta in the female

1 Thymus

1 Pineal gland or body.

Pituitary Gland (Hypophysis)

The human pituitary is a reddish-gray oval structure, about 10 mm in diameter, located in the brain just behind the optic chiasma as an extension from the floor or the hypothalamus. It is lodged in a depression, the *sella* turcica or hypophyseal fossa of the sphenoid bone. In man it weighs about 0.5 gm.

Structure: The pituitary gland has a dual origin, both components being derived from the *ectoderm.* These are adenohypophysis and neurohypophysis. The adenohypophysis arises from the roof of the buccal cavity in the form of an outgrowth called "Rathke's pouch". The neurohypophysis originates as a downgrowth from the floor of the diencephalon in the fore brain. Adenohypophysis soon loses its connection with the buccal cavity whereas neuronypophysis remains attached to the brain by an *infundibula or pituitary stalk.*

The Adenohypophysis: This is composed of three parts– the pars intermedia, the pars distalis and, the pars tuberailis (the two latter being the anterior lobe' of the older terminology). The adenohypophysis consists of cells (chromosomes and chromophils) which themselves secrete with the exception of the pars tuberalis which is simply a supporting structure for the gland and its blood vessels.

The Neurohypophysis: It comprises Infundlbulum or Neural Stalk and pars nervosa or neural lobe. The neural lobe (Infundibular process) consists of non-myelinated nerve fibres and non-secretary supporting cells. The hormones of neurohypophysis are now clearly established to be formed in the supra-optic and paraventricular nuclei of the hypothalamus of brain. Two tracts of non-medullated nerve fibres arise from these nuclei and supply the posterior lobe. These tracts contain motor fibres only. It is believed that these hormones are released from the nerve terminals which contain them when impulses pass down the axon, which means that the hypothalamic neurons concerned are both nervous and secretory in function.

The production of hormones by nerve cells is known as neuro-secretion. Thus, the neural lobe or infundibular process itself is simply a storage depot for hypothalamic secretions and does not contain its own secretory cells.

The blood supply to the pituitary is by the superior and inferior hypophyseal arteries which are branches from the internal carotid arteries.

Venous drainage is into one of the venous sinuses of the brain.

Functions: The pituitary gland is often described as the *master gland* of the endocrine system because it secretes a number of hormones which regulate and control the activity of other endocrine glands.

Anterior Pituitary or Adenohypophysis Hormones

(I) Growth Hormone (GH) or Somatotropic Hormone, (STH) : This hormone stimulates growth of bones, cartilage, muscles, viscera and the body as a whole. Growth hormone promotes protein anabolism, the absorption of calcium from the intestine and the conversion of glycogena to glucose. (Growth hormone has direct actions on cellular activity and metabolism).

Hyposecretion

(a) *During Childhood:* Due to lack of the growth hormone, there is stunted growth of the skeleton resulting in dwarfism. There are two types of dwarfs:

(i) *The Lorain Type:* A minute individual, fairly well

proportional who is mentally and physically normal.

(ii) *Frohlich's Type:* In this type of dwarfism growth is stunted. There is considerable obesity, mental deficiency and retarted sexual development.

(b) *During Adult Life:* This condition is very rare and is probably due to atrophy of the anterior lobe of the gland. This condition is exhibited by early *senility*. The skin becomes dry and wrinkled, the hair grey and sparse, the body emaciated, and there is degeneration of the sexual organs with cessation of menstruation in the female. This condition is known as Simmond's disease.

Hypersecretion

(a) *During Childhood:* Due to increased secretion of the growth hormone gigantism occurs. There is excessive skeletal growth before the epiphyses have closed and the individual may becomes 8 or 9 feet tall.

(b) *During Adult Life:* The hyperactive gland in the adult (after union of epiphyses of bones) results in a condition known as acromegaly. There is excessive growth of the bones of the face, specially the frontal bone and the mandible. The hands and feet become large and spade-like. There is also thickening of the skin on the face and hands.

The Pituitary Tropins: Pituitary hormones which influence the activities of other endocrine glands (*e.g.* Adrenal cortex, thyroid and Gonads), principally those involving reproduction or stress; are called tropic hormones. They are carried by the blood to other target glands and aid in maintaining these glands and stimulating production of their respective hormones.

Adrenocorticotropic Hormone (ACTH): This hormone stimulates the *cortex* of the *adrenal gland* to produce Glucocorticoids Production of CTH is particularly important during emotional and physical stress.

Thyroid Stimulating Hormone (TSH) or Thyrotropic Hormone (TTH): This hormone controls the growth and activity of the *thyroid gland.* It influences the uptake of iodine, the synthesis of thyroid hormones (T_3 and T_4) by the thyroid gland and the release of stored hormones into the blood stream.

Gonadotropic Hormones: These regulate the development and functions of the gonads.

Follicle Stimulating Hormone (FSH)

(a) *In the Female:* The target organs are the ovaries where FSH stimulates the development and ripening of the ovarian follicle. During its development the ovarian follicle secretes its own hormone *estrogen.*

(b) *In the Male:* It stimulates the development of seminiferous tubules and spermatogenesis.

Due to it's action on both male and female gametes, FSH is also called Gametokinetic factor.

Luteinising Hormone (LH)

(a) *In the Female:* This hormone promotes the final maturation of the ovarian follicle, ovulation and formation of the *corpus luteum* which secretes the second ovarian hormone known as *progesterone.*

(b) *In the Male:* It stimulates the interstitial cells of testes causing them to release male sex hormones (androgens) into the blood stream. Hence, in the male, it's also known as *Interstitial Cell Stimulating Hormone (ICSH).*

Feedback Control of Tropic Hormones: The tropic hormone are usually subject to Feedback inhibition at the pituitary or hypothalamic level by the hormone product of the final target gland. Thus when the blood level of Adrenocorticosteroids, sex *steroids* and thyroid hormones becomes high they inhibit the release of their respective tropic hormones. If the concentration of these hormones in the blood falls, the hypothalamus and the anterior lobe of pituitary secrete more of the tropic hormones. Through such control systems the output of tropic pituitary hormones is automatically self-adjusting. The blood level of the hormones is thus maintained and their production related to needs. This type of regulatory system of hormonal secretion is called "feedback", "plus-minus" or "pull and push" control mechanism that operates between the hormone and the target organ.

Lactogenic Hormone (Prolactin, PL, Mammotropin, Leuteotropic Hormone LH): The lactogenic hormone has a direct effect upon the *breasts* or *mammary glands* immediately after the delivery of a baby and the expulsion of the placenta. In conjunction with other hormones like growth hormone, ACTH and TSH, prolactin stimulates the breasts to secrete milk. During pregnancy, the ovarian hormones inhibit the secretion of prolactin. Besides, prolactin induces maternalism, strong emotional attachment to the young and drive to bring the offspring to a state of self-sufficiency. In males, prolactin promotes protective attitudes towards mates and offspring, intensive food gathering for the family and expressions of Paternalism in general.

Melanocyte Stimulating Hormone (MSH) or Intermedin: MSH is produced in the pars intermedia. Though found in all vertebrate groups, this hormone appears to be functional only in poikilothermic ones. Its secretion causes a dramatic darkening of the skin of many fishes, amphibians and reptiles. The darkening occurs because of the spread of granules of *melanin* through the branches of specialised *melanocyte cells* in the epidermis. When the melanin is concentrated in the centre of the melanocytes, the skin has a light appearance. This mechanism is of great value in enabling the animal to blend in with its surroundings.

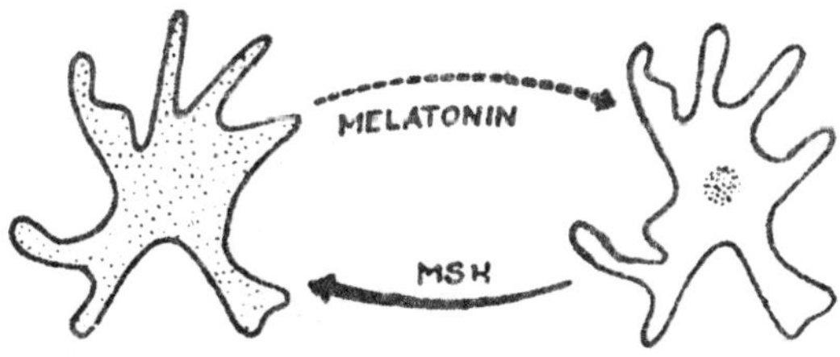

A generalised vertebrate melanocyte. Dispersed condition (Skin dark) Concentrated Condition (Skin Light).

Although MSH does not seem to play an important role in the normal behaviour of human melanocytes, but under certain conditions such as pregnancy, an increase in its secretion does cause some darkening of the skin. Normally, large number of melanocytes are present in the human skin, where they are responsible for moles and freckles.

Posterior Pituitary or Neurohypophysis Hormones: Whereas adenophypophysis secretes hormones on its own, neurohypophysis is not secretory in function. It consists of non-myelinated nerve fibres and supporting cells and receives its quota of hormones from neurosecretory cells in the hypothalamus of brain.

(i) *Infundibulum:* This part sends a number of release factors or hormones that travel by portal circulation to the anterior pituitary and trigger the secretion of a series of tropic hormones:

Release Factors

No.	*Hypothalamic Hormone or Release Factor*	*Abbreviation*
1.	Corticotropin (ACTH)—release factor	CRF
2.	Thyrotropin (TSH)—release factor	TRF
3.	Luteinising hormone (LH)—release factor	LH-RF
4.	Follicle-stimulating hormone (FSH)—release factor	FSH-RF
5.	Growth hormone (GH)—release factor	GH-RF
6.	Prolactin—release factor	PRF
7.	Prolactin release—inhibiting factor	PIF
8.	Melanocyte—stimulating hormone (MSH) release—inhibiting factor	MIF
9.	Somatostatin or GH Inhibiting factor	GHIF

These release factors (hormones) may be built up of 3 to 20 amino acids units. For example, TRF is a tripeptide, whereas GH-RF is a decapeptide. The study of these release factors was considered so important that the Noble prize in physiology or

Medicine for 1977 has been awarded to Andrew V. Schally and Roger C.L. Cuillemin in U.S.A. for isolating, synthesising and characterising the polypeptides from the hypothalamus which control the anterior pituitary function.

(ii) *The Infundibular Process:* Extract of the posterior lobe is called Pituitrin which contains 2 hormones:

Oxytocin (Pitocin) ("Birth Hormone" or "Milk Ejection Hormones"): *(i)* Produces *contraction of the uterine muscles (a)* to facilitate the ascent of spermatozoa after coitus and (b) for delivery of the foetus. In late pregnancy, the uterus becomes very sensitive to oxytocin. The amount is increased during labour, *(ii)* Facilitates the *flow of milk* from mammary glands, by contraction of the myoepithelial cells of the lactating breast, squeezing milk into the large ducts behind the nipple. Sucking plays an essential part in reflex stimulation of the pituitary gland. When sucking stops location will continue only for a short time.

Antidiuretic Hormone (ADH), Vasopressin or Pitressin: This has two important functions:

(i) Influences water-balance of the animal by reducing its output of urine. It exerts a direct effect upon the facultative reabsorption of water by the kidney tubules.

ADH may be increased by a variety of stimulators of neural activity. Emotional and physical stress, electrical stimulation, acetylcholine, nicotine, and morphine increase ADH secretion.

If the antidiuretic hormone is lacking, reabsorption of water is reduced and large amounts of dilute urine are excreted. This ailment is called *diabetes insipidus.* It differs from the more common *Diabetes mellitus* in that the urine is free from sugar *(mellitus* is 'sweet' diabetes, while *insipidus* is 'tasteless' diabetes, the common feature in both being the excessive flow or "syphoning off or urine).

(ii) Under the influence of ADH, the involuntary smooth muscle of the intestine, gall bladder, urinary bladder and blood vessel contracts. Contraction of the muscle layer in the blood vessel walls raise the blood pressure.

Thyroid as a Source of Hormone

It is a bilobed gland situated in the lower part of the neck, ventral to the trachea and immediately behind the larynx. The two lobs of the thyroid are joined together by a narrow bridge of tissue called *isthumus* which gives the entire gland a more or less H-shaped appearance.

Histologically, it consists of *acini* or *vesicles* lined with cuboidal epithelium. The lumen of these acini is filled with a proteinaceous colloidal secretion, *thyroglobulin* (an iodised glycoprotein).

Synthesis of Thyroid Hormones: The thyroid gland produces hormones which are iodinated derivates of the amino acid throsine. The formation of these hormones takes place in several stages:

(1) The first stage in their production is the accumulation of *inorganic iodide* in the aceiner cells of the thyroid from the blood plasma. This accumulation is accelerated by the *thyrotropic hormone (TSH)* of the pituitary.

(2) As iodide is absorbed by the aciner cells, it is oxidised enzymatically to iodine $2I^- == I_2 + 2e$.

(3) The iodine is immediately taken up by the amino-acid tyrosine of the gland cells to form *Mono* - and *Di-iodotyrosine (I Tyr I_2 Tyr).*

(4) These then undergo *oxidative coupling (condensation)* to form *derivatives of Thyronine.* One Mono-and one Di-iodotyrosine units combine to form a *Tri-iodothyronine (I_2 Thn),* whereas two Di-iodotyrosine units form Tetra-iodothyronine (I_4 Thn) which is also referred to as *Thyroxine.* It was first crystallised by Kendall in 1919.

(5) The Thyroxine and l_3 *Thn* produced in the gland are

still in peptide linkage within the thyroglobulin molecule and are stored as such in the colloid.

(6) Under simulation by TSH the thyroglobulin is broken down by a *hydrolytic protease enzyme* to yield *Thyroxine* (80 per cent) and I_3 *Thn.* (20 per cent) which pass into the blood. The enzyme is secreted by the aciner cells and derived from their lysosomes.

(7) In the blood, I_4 Thn and I_3 Thn are transported in association with two plasma proteins — *thyroxine-binding globulin* and *thyroxine binding prealbumin*. These act as specific carrier agents for the hormones and represent the Protein Bound iodine (PBI) of the plasma.

(8) Finally, they enter the cells of tissues where about 50 per cent of the *Thyroxine* is dehalogenated (one I_2 removed) by a dehalogenating enzyme to I_3 *Thn* and *inorganic iodide* which is returned to the thyroid gland.

(9) Both I_4 *Thn* (65 - 90 per cent) and I_3 *Thn* (10 - 35 per cent) are active hormones but I_3 Thn is 3-5 times more active than I_4 *Thn* and has a more rapid onset of action.

Functions of Thyroid Hormones

(1) An adequate supply is essential for normal development and growth of body.

(2) They are responsible for the maintenance of healthy hair and skin.

(3) They accelerate O_2 consumption by the cells, energy production and BMR (Calorigenic action).

(4) They influence heat production during the ketabolism of food-stuff in the cells.

(5) They increase the intestinal absorption of glucose ; glycogenolysis, deamination and gluconeogenesis in liver cells.

(6) They enhance rate of heart-beat and blood pressure.

(7) They are associated with nerve stability controlling the excitability of nerve fibres.

(8) They are essential for metamorphosis in amphibia.

(9) They regulate reproduction and cellular physiology.

Abnormalities of Thyroid Function

Hyposecretion: A deficiency of thyroid hormones produces a number of clinical disorders depending upon the degree of the deficiency and the age at which it occurs:

(a) In Infancy: Cretinism

(1) Lack of skeletal development resulting in deformed bones and stunted body growth.

(2) Mental deficiency due to subnormal development of the nervous system.

(3) There is a large protruding tongue from a sagging mouth and pot-bellied abdomen.

(4) The BMR is lowered accompanied by slow pulse and respiration rate.

(5) Sex organs and secondary sexual characters becomes retarded.

(b) In Adult Life: Myxoedema (Gull's Disease)

(1) The BMR, body temperature and pulse rate are lowered, and there is undue sensitivity to cold.

(2) There is considerable general oedema particularly manifested by the puffiness of face and hands due to thickening of the skin by the deposit of an albuminium semi-fluid material underneath.

(3) The skin becomes coarse and dry and there is loss of hair.

(4) Anaemia and slowing of physical and mental reactions.

(5) Sexual development is retailed.

Simple (Endemic or Colloid) Goitre: It is a deficiency disease caused by an inadequate supply of iodine in the diet. Simple

Goitre is fairly prevalent in certain countries such as Switzerland, the Himalayan regions and in the Pyrenees, where the iodine content of water and soil is low. In this disease there is a pathological enlargement of the gland itself, which protrudes, as a marked swelling in the neck. This enlargement is believed to be due to overestimation of the gland by the thyrotropic hormone from the anterior pituitary in the absence of "braking" or feedback inhibition of the thyroid hormone.

Since only small amounts of iodine are necessary for the formation of thyroid hormones, treatment with iodine or with iodised salt (NaI_2) is adequate for cases of simple goitre.

Hypersecretion: This condition generally occurs in adult life, and is also referred to as *thyrotoxicosis* or *hyperthyroidism.*

Exophthalmic Goitre or Graves' Disease or Basedow's Disease (Plummer's disease): The chief symptoms are:

(1) The BMR is greatly increased and the individual appears to be literally "burning himself up".

(2) There is a raised nervous excitability, restlessness and anxiety, accompanied by fatigue, muscular weakness, tremors and loss of weight.

(3) There is an increase in the heart rate and pulse rate with a corresponding increase in respiration rate and copius sweating.

(4) Increased body temperature with excessive sweating.

(5) A characteristic protrusion of the eye-balls (exophthalmus) with a staring look, usually accompanies hyperthyroidism. This protrusion is thought to be due to increased post-orbital pressure and possible weakening of the extrinsic muscles of the eye, as a result of excessive TSH.

Hyperthyroidism can be treated by thyroid removed by *surgery* or by *radioactive iodine* (^{131}I). It is also treated by *anti-thyroid drugs (goitrogens).*

Antithyroid Drugs: These are used experimentally to produce hypothyroidism (as in the amphibian metamorphosis), and clinically in the treatment of hyperthyroid conditions. These inhibit the production of thyroxine both at the organification and at the coupling steps.

Examples of these anti-thyroid drugs are goitrogens—thiouracil, propylthiouracil, methylthiouracil carbimazole, thiourea, and methimazole.

Thyrocalcitonin (TCT) is a non-iodinised calcium-lowering hormone originating from the thyroid gland. The release of TCT is stimulated by high levels of ionised calcium in the serum. The hormone lowers the level of calcium in blood and increases calcium excretion by the kidneys.

Significance of Parathyroids

The parathyroids are four small glands placed on the thyroid, two on each lobe. In man, the glands are reddish or yellowish-brown egg shaped bodies; the 4 glands together weigh about 0.1 to 0.2 g. The glands were first recognised by Raynard (1835) and their structure was described by Sandstrom (1880).

Hormone of the Glands: They secrete a hormone, called *Paratharmone* or *Parathyroid hormone (PTH)* OR *Collip's Hormone* after its discoverer Phillip Collip (1925). It is under the feed-back control of blood Ca^{++} level.

Functions

(1) The primary function of paratharmone is to maintain the concentration of ionised calcium (Ca^{2+}) in the plasma, within the narrow range characteristic of this electrolyte, despite wide variations in calcium intake, excretion, and deposition in bone.

(2) Lowers the serum phosphorus by increasing urinary excretion of phosphate but decreases excretion of calcium.

(3) Increases the concentration of Ca^{++} in the blood plasma by mobilising calcium from bones, particularly if the dietary intake of calcium is inadequate.

(4) Enhances the rate of absorption of calcium from the intestine.

The growth of bones, muscle-tone and normal-nervous activity is dependent on a fixed Ca-P balance. Normally in man there is 12 mg of Ca and 4.5 mg of inorganic phosphate per 100 ml of blood.

Secretion of PTH is subject to control by a negative feedback mechanism relating to the levels of ionised Ca^{++} in the plasma.

Abnormalities of Parathyroid Activity

Hyposecretion: Removal of the parathyroid glands in error during the operation of thyroidectomy, or due to some local condition of the glands themselves causes a decrease in the secretion of paratharmone. In this state the level of free calcium in the blood is reduced, because much of the calcium in the blood is bound to other substances. As a result the muscles go into a type of spasm known as tetany (sustained contractions of the muscles of larynx, face, hands and feet).

Characteristic of Tetany:

(1) Muscle twitching leading to muscle spasm of the hands and feet termed carpo-pedal spasm.

(2) Increased neuromuscular irritability and spasm of the eye muscles.

(3) In infants and children spasm of the muscles of the larynx occurs accompanied by cyanosis.

(4) Serum level of calcium falls by more than 50 per cent and phosphorus shoots high, resulting in abnormally greater excitation of the nerve fibres and muscles.

Hypoparathyroidism occurs only very rarely.

Hypersecretion: This is rare and is probably due to a tumour of one mere of the parathyroid glands. In case of hypersecretion of Paratharmone osteoclasts (the cells which destroy bone) become overactive so that there is an excessive destruction of bone and non-utilisation of calcium. This results in softening of the bones and a high blood calcium. Eventually calcium is deposited in the walls of arteries and lungs. In the areas of destruction in the bones fibrous cysts develop. This condition is called osteitis rosa cystica. The bones becomes painful and fractures occur frequently.

Adrenal or Suprarenal Glands

The adrenal glands are situated one on each side of the vertebral column, closely applied to the upper poles of the kidneys ; hence the name supra-renals. The adrenal glands are surrounded by a capsule of areolar tissue containing fat.

Structure: The glands are composed of two distinct parts which differ both anatomically and physiologically. The outer part (derived from mesoderm) is known as the cortex and the inner part (derived from neural crest) as the medulla.

Cortex: It is composed of three layers of cells:

(i) The outer layer zona glomerulosa, is made up of comparatively small thickly set ovoid groups of cells. These secrete mineralocorticoids which regulate salt and water balance.

(ii) The middle layer, zona fasciculata is the widest layer which is made up of polyhedral cells arranged in radiating columns. This layer secretes glucocorticoids affecting carbohydrate metabolism.

(iii) The inner layer, zona reticularis is made up of an irregular network of rows of cells. These secrete sex hormones.

Medulla: The medulla is completely surrounded by the cortex. In man this region is thin, comprising only 5-10 per cent of the gland. The cells of the medulla assemble in rounded groups or short anastomosing columns with many intermingling blood capillaries. The medulla is composed of sympathetic gang-lion cells which have no post-ganglionic fibres. Medullary cells are essentially neurosecretory cells because they are stimulated by pre-ganglionic sympathetic fibres to produce the hormones *adrenatine* and *moradrenaline.*

Medullary cells have a special affinity for chromium compounds such as $K_2Cr_2O_7$ and Chromic acid, and they are, therefore, also referred to as *chromaffine cells.*

Functions

Cortex: The adrenal cortex produced a number of potent hormones all of which are steroid derivatives:

The Mineralocorticoids: These are associated with the maintenance of electrolyte balance in the body and distribution of water in tissues.

Specific examples are *aldosterone* and *deoxycorticosterone.* The most potent mineralocorticoid is aldosterone. Deoxycorticosterone is only 4 per cent as potent as aldosterone.

Functions of Aldosterone

(1) It is a "salt retaining hormone" which promotes the reabsorption of sodium and chloride ions by renal tubules. Retention of these in the blood keeps its osmotic pressure high, as a result of which water is also reabsorbed. This in turn assures normal blood volume and pressure.

(2) Increases excretion of potassium ions.

The amount of aldosterone produced is influenced by the sodium level in blood. When the sodium blood level falls, more aldosterone is secreted and more sodium reabsorbed.

Aldosterone secretion is stimulating by the angiotensin produced by kidneys.

Hyposecretion of aldosterone causes Addison's disease.

This disease was first described by Thomas Addison *(Father of Endocrinology)* in 1855. It is characterised by:

(1) Excessive loss of sodium chloride in the urine and elevated levels of potassium in the serum.

(2) Lowering of blood pressure, body temperature and BMR.

(3) Muscular weakness.

(4) Hypoglycemia.

(5) Loss of appetite, vomiting and gastro-intestinal disturbances.

(6) Brownish pigmentation or bronzing of the skin especially the exposed parts, for example, the face and hands due to deposition of melanin.

The Glucocorticoids: These primarily affect metabolism of carbohydrate, protein and lipids.

Specific examples are *corticosterone, hydrocortisone* or *cortisol* and *cortisone*. Their secretion is stimulated by ACTH from the anterior lobe of pituitary.

Functions

(1) They are metabolically antagonistic to insulin. Under their influence glycogen is changed to glucose in the liver, raising the blood sugar level (hyperglycemic action).

(2) During stress, they promote the conversion of fats and proteins to glycogen that is ultimately converted to glucose (gluconeogenesis).

(3) They have anti-inflammatory and anti-allergic properties. Cortisol is used in the treatment of rheumatoid arthritis when it is even injected inside of the diseased joint. It is a life saving drug in anaphylactic or adverse reactions to certain drugs- like reaction to the injection of penicillin and in severe attacks of asthma.

This property of cortiocosteroids is called immunosuppressive whereby they suppress the normal immune reactions of body against antigens and attack of micro-organisms. For this very reason, cortisol in often used for treatment of allergy. It is also useful in the form of ointments for skin and eye diseases.

(4) In recent years, this hormone, has been utilised to suppress the phenomenon of rejection in transplantation of organs.

(5) Together with ACTH, the glucocorticoids serve to counteract the harmful effects of long-lasting stress such as starvation, cold, heat, burns, toxins, infections, forced exercises, etc.; by helping the tissues to develop resistance and adapt them to meet new conditions. The endocrinilogist Hans Selye calls these responses "the general adaptation syndrome".

Hypersecretion of cortisol in adults results from one of the two causes:

(1) Adrenocortical hyperplasia initiated by increased production of ACTH due to hyperactivity of basophil cells in the pituitary gland.

(2) Benign or malignant tumours of the adrenal cortex. The term *Cushing's disease* has been restricted to those cases which are of pituitary origin whereas *Cushing's Syndrome* denotes adrenocortical hyperfunction directly involving the adrenal glands.

The signs of hyperadrenocorticism (Cushing's disease and Syndrome, 1932) include:

(1) Hyperglycemia and glycosuria (diabetogenic effect).

(2) Excessive deposition of fat over the face, back of the neck and abdomen.

(3) Muscular weakness due to potassium depletion through the kidneys.

(4) Wasting of the muscles due to protein anti-anabolic effect and glyconeogenesis.

(5) (a) *Virilism* in the female, *i.e.,* a tendency to develop male characteristics, viz., (i) there is excessive growth of hair on the chest and public region with an increase and darkening of the facial hair (hirsutism) (ii) cessation of the menstrual flow (amenorrhoea) and (iii) gradual ·atrophy of the breasts (iv) overgrowth of clitoris, (b) *Feminism* in the male, *i.e.,* a tendency to develop female sex characteristics, *i.e.,* enlarged mammary glands (gynaecomastia) and retailed growth of bread.

The Anrogens or Estrogens: These primarily affect secondary sex characteristics in their specific target organs.

The primary adrenal androgens are *Dehydroepiandrosterone (DHEA) and Androstenedione.* These cause retention of nitrogen (a Protein-anabolic effect), phosphorus, potassium, sodium and cloride. If present in excessive amounts, they also lead to masculinisation in the female. Adrenal estrogens influence the development and maintenance of the secondary sex characters in females.

Medulla and its Hormone

The adrenal medulla is a derivative of the symphathetic part of the autonomic nervous system. When stimulated by pre-ganglionic symphathetic nerve fibres, it secretes two hormones: *adrenaline (epinephrine)* (80 per cent) and *norarenaline (norepinephrine)* (20 per cent). These hormones belong to a class of substances known as catecholamines that are derivatives of amino acid tyrosine.

Functions of Noradrenaline: It is a precursor of adrenaline which does not possess a methyl group (—CH_3), as is found in the adrenaline. Nor adrenaline produces a vasoconstriction of peripheral organs like skin ; digestive and reproductive organs. This effect raises blood pressure.

Functions of Adrenaline:

(1) Causes vasodilation of the arterioles of the skeletal muscles increasing the supply of oxygen and nutritional materials for a sustained muscle activity.

(2) Causes vasoconstriction of the arterioles of the skin, mucous membranes and splanchnic viscera, directing the blood away from them to the skeletal muscles, brain, coronary artery and liver.

(3) Dilates trachea and bronchi to permit a greater volume of air to enter the lungs at each inspiration.

(4) Increases the rate of heart beat, thus increasing blood pressure.

(5) Slows down peristalsis in the digestive tract and limits the flow of saliva.

(6) Increases the tone of the anal and urethral sphincter muscles, thus inhibiting micturition and defecation.

(7) Stimulates the conversion of liver glycogen stored in liver and muscles to glucose that is ready for oxidation.

(8) Decreases *insulin,* thereby preventing the glucose from being taken up by peripheral tissues and preserving it for the CNS.

The net of such changes is that an increased quantity of blood is pumped faster to heart, muscles and brain ; it is well oxygenated and rich in nutrients (stress or alarm reactions). These preparations are valuable for the body in providing a rapid physiologic response to emergencies such as fatigue, shock, fear, excitement and danger. In this sense, it mobilises what has been termed the "fight or flight" mechanism, *i.e.,* the emergency function, as suggested by Cannon.

Pancreas and its Role in Digestion

The pancreas is a rather diffuse gland which is a mixture of two histologically and functionally separate tissues. The bulk of the gland is exocrine in function and forms clusters of glandular cells called acini or lobules lined with cuboidal epithelium. Scattered among the exocrine cells are patches of tissues named Islets of Langerhans after their discover and these consists of endocrine cells that discharge their secretion directly in the blood. These cells are small, prism-shaped, closely packed together and without a lumen ; they collectively form an endocrine gland. In mammals, two distinct types of cells are recognisable in the islet tissue, the α-cells which contain alcohol-soluble granules, and the β-cells, whose granules are not soluble in alcohol. These two cell types each secrete a different hormone, *glucagon* being secreted by α-cells and *insulin* by the β-cells.

Role of Insulin

Insulin is the hypoglycemic anti-diabetic factor and the protein hormone which regulates blood glucose.

History: Banting and Best extracted insulin in the year 1921 for which they were awarded Noble Prize in 1923. Abert prepared pure crystalline insulin in 1926.

Chemical Structure: Insulin is a relatively small protein of 51 amino acids disposed in two parallel chains A and B, interconnected by two disulfide bonds. A chain has 21, and the B chain 30 amino acid units ; thus one molecule of Insulin has 51 amino acids in all. Its molecular weight is 6,000.

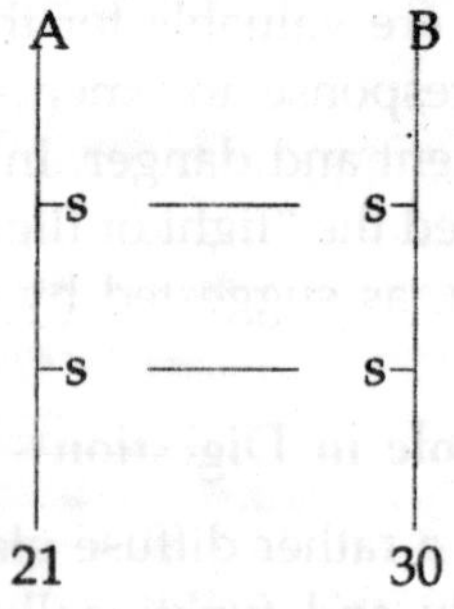

The Insulin Structure

Functions

1. *Increases Combustion of Sugar in tissues:* Insulin accelerates the phosphorylation of glucose by the enzyme Glucokinase or Hexokinase. Glucose-6- phosphate thus formed enters the respiratory cycle to liberated chemical energy.

   $$\text{Glucose} \xrightarrow[\text{Glucokinase}]{\text{Insulin}} \text{Glucose-6-phosphate}$$

2. It helps in the *transport of glucose* into the cell.
3. *Stimulates Glycogenesis:* Insulin promotes the formation of glycogen from glucose in the liver and muscles.
4. *Prevents Gluconeogenesis:* It reduces the production of glucose from non-carbohydrate sources such as proteins and fats.
5. *Prevents formation of Ketone Bodies (Antiketogenic Effect):* When the supply of insulin is inadequate fat metabolism is also affected, because fat must be used instead of carbohydrate for the release of energy, through cellular oxidation. The large increase in the amount of $CH_3CO.CoA$ which accumulates from the breakdown of fat is too great to be dealt with by the cellular TCA Cycle. Instead of being combined with oxaloacetic acid in the first stage of that cycle, it passes to the liver where it is condensed to form acetoacetic acid (a 4C) molecule). The liver is unable to make use of acetoacetic acid ; and, therefore, the aid is delivered to the body tissues by circulation. In the blood, the acetoacetic acid undergoes further dissimilation forming β *-Hydroxybutyric acid* and *acetone*. These degradation products are known as Ketone bodies.

   Acetyl CoA
   (2C)
   ↓

6. *Stimulates protein-synthesis and growth, e.g.,* nitrogen-

retension, bone formation, etc. Increases the incorporation of amino acids into peptides.

7. *Decreases Pyaemia, i.e.,* accumulation of excess fat in blood and liver.

Significance of Diabetes Mellitus

Insulin deficiency causes diabetes mellitus. This is a disorder of carbohydrate metabolism characterised by:

(1) *Hyperglycemia:* It is a condition in which blood sugar increases above the mammal level (80 to 120 mg/ 100 ml).

(2) *Glycosuria:* It is a condition in which glucose is present in the urine in such quantities as will reduce Benedict's or Fehling's reagent. It occurs when the blood sugar level exceeds the renal threshold which is 180 mg.

(3) *Polyuria:* The high glucose level in the nephric filtrate increased osmotic pressure, sharply reducing the reabsorption of water back into the blood, hence victims of the disease urinate large volumes of urine frequently. This is known as polyuria.

(4) *Dehydration:* The continued loss of water and electrolytes in the urine leads to increasing dehydration.

(5) *Polydipsia:* Because of polyuria, the patient is thirsty inspite of drinking large amounts of fluid.

(6) *Weight Loss:* Inspite of *polyphagia* (increased appetite) the patient loses weight because of the excessive breakdown of protein.

(7) *Abnormal Metabolism of Fat:* Muscles are unable to utilise the glucose in the absence of Insulin, therefore the diabetic feels weak and tried. Since carbohydrate cannot be used as fuel, fat is used instead to provide the energy.

(i) *Lipaemia:* Fat is mobilised from the body stores and

transferred to the liver, hence fat content of the liver and blood rises.

(ii) *Ketosis:* The disproportionate metabolism of fat in diabetes results in the overproduction of ketone bodies.

(iii) *Acidaemia or Acidosis:* As acetoacetic acid and P-hydroxybutyric acids are produced faster than they can be metabolised, the patient develops acidosis.

(8) *Diabetic Coma:* The keto acidosis is associated with increasing drowsiness and if untreated, the patient may become unconscious (diabetic coma) and die from a combination of hyperglycemia, ketosis, acidaemia and dehydration.

(9) Reduced Resistance to infections due to altered immunological response and altered metabolism, etc.

Stimulus to Insulin Secretion

It is a high blood sugar level. There are no secretory nerve fibres to the islets ; when the blood sugar rises, this rise, is all the stimulus the islets need to increase their output.

Therapy: Insulin is now available in large quantities from the glands of slaughtered livestock (cattle an pigs). The molecular structure of these animal insulins is so close to that of human insulin that they may be used to treat diabetes. Insulin is protein in nature therefore, it is inactive when given by mouth, since it is destroyed by the digestive enzymes.

In recent year some types of "modified", insulin have appeared. Some are combined with Protamine, which is another protein and some with zinc salts:

1. IZS (Insulin Zinc Suspension)
2. PZI (Protamine Zinc Insulin)

The purpose of such modification is to render the insulin less soluble so that a given dose will be more slowly absorbed.

Insulin Reaction or Shock: Insulin injected after a period of exercise or long after a meal may drive the blood sugar level down to abnormally low levels, 43 mg 100ml. The victim becomes irritable, fatigued, and may lose consciousness. This condition is of insulin shock.

Glucagon and its Utility

It is a polypeptide consisting of 29 amino acids arranged in a single chain. The effect of glucagon is to cause a rise in the blood glucose level by mobilising the glycogen in the liver, which is broken down to glucose.

The pancreatic hormones are secreted into the hepatic portal vein, so that the liver forms a barrier between the hormones and the body cells.

The liver destroys glucagon before it can get into the body circulation, and its metabolic action is exerted solely upon the liver cells. Therefore, the ability of the liver to destroy the hormone has resulted in failure of its effect on muscle glycogen.

Role of Alimentary Canal

Historically, some of the hormones concerned with the digestion of food were among the first hormones to be discovered. They are 'local' hormones, being produced by a localised area of the gut and acting upon the gut itself or one of its accessory glands.

Stomach

Gastrin: Gastric distension and protein in the stomach stimulates the gastric mucosa to secrete a hormone gastrin, which has a strong stimulating effect on the acid-secreting parietal cells of the stomach.

Duodenum: The duodenal wall secretes several hormones when food is passed into it from the stomach.

(i) *Secretin:* It was the first hormone to be discovered (in

1902) which circulates in the blood, and when it reaches the pancreas, it causes the exocrine cells of the gland to secrete the fluid base of the pancreatic juice (water and electrolytes like Na^+ HCO_3 and Cl^- with smaller concentration of K^+ and Ca^+).

(ii) *Pancreozymin (PZ):* A second duodenal hormone, pancreozymin, follows the same path to evoke the production of the pancreatic enzymes.

(iii) *Cholecystokinin (CCK):* Stimulates the gall bladder to rhythmic contractions and these result in an increase in the rate of flow of bile into the duodenum.

Small Intestine

It secretes two hormones:

(i) *Enterocrinin:* The hormone stimulates the intestinal mucosa to secrete the intestinal juice.

(ii) *Enterogastrone:* The presence of fats in the small intestine stimulates the intestinal mucosa to produce another hormone *enterogastrone* which eventually stops the secretion of gastric juice in the stomach and cairns down peristalsis there.

Kidney: Erythropoietin is formed by kidney that acts on haemopoietic organs for the stimulation of the formation of blood cells, especially in response to anaemia. Erythropoietin acts on bone marrow to increase its production of R.B.C's.

Gonads: The testes and ovaries, in addition to their cytogenic function of producing spermatozoa or ova, manufacture steroid hormones which control secondary sex characteristics, the reproductive cycle, and the growth and development of the accessory reproductive organs.

Male Hormones: In mammalian testes, the Interstitial (Leydig) cells lying scattered in between the seminiferous tubules synthesise male hormones, the androgens. The principal male hormone is *testosterone,* which is synthesised from cholesterol through pregnenolone, progesterone, and hydroxyprogestrone. The latter is then converted to the C-19 ketosteroid-androstenedione, the immediate precursor of testosterone.

Of the synthetic products, *methyl testosterone* and *testosterone prepionate* are important members.

The principal metabolites of testosterone are *androsterone* and *etiocholanolone,* the major 17-ketosteroids in the urine.

Functions of Testosterone

(1) Promotes the growth and functions of the epididymis, vas deferens, prostate, seminal vesicles and penis.

(2) Stimulates the development of male secondary sex characters — (i) the voice deepens (ii) hair begins to grow on the face and on the upper lip ; in the axillae and in the pubic region. (iii) there is a greater skeletal and muscular growth, (iv) aggressive nature and sex urge. In brief, it determines libido.

(3) Influences spermatogenesis and seminal plasma secretions.

(4) Exerts a feedback inhibitory effect on pituitary LH secretion.

(5) It has a general anabolic effect on the body by stimulating protein synthesis.

Female Hormones: Two main types of female hormones are secreted by the ovary: the *estrogens* or *follicular hormones* produced by the cells of developing graafian follicle and the *progesterone* derived from the corpus luteum that is formed in the ovary from the ruptured follicle ovulation.

Function of Estrogens: The principal estrogenic hormone in the blood circulation is *estradiol* which is secreted by the internal cells of the graafian follicle. The estradiol is excreted as *estrone* or *estriole.*

(1) Promotes the development of female secondary sex characteristics. At the start of the adolescence (puberty),

   (i) The mammary glands (breasts) develop chiefly by proliferation of ducts.

   (ii) The pelvis becomes broader,

   (iii) Voice stays high pitched,

   (iv) The genitalia enlarge and there is growth of hair around them and in the axillae,

   (v) Development of fat deposits in subcutaneous tissues and other particular regions to make a typical feminine body (characterised by rounded body contours).

(2) Prepares the uterine mucosa for the later action of progesterone, *i.e.*, proliferative growth of the lining of the endometrium, deepening of uterine glands, and increased vascularity.

(3) Influences growth of fallopian tubes and stratification of vaginal epithelium.

(4) Suppresses the production of pituitary FSH.

(5) Influences behaviour and psychic patterns in females.

Functions of Progesterone: After ovulation, the discharged follicle cells become transformed into a characteristic body known the corpus luteum, a temporary endocrine structure that secretes the hormone, *progesterone.* It is a steroid which is synthesised from its immediate precursor, pregnenolone.

(1) Causes extensive development of the endometrium, preparing the uterus for the reception of the embryo and for its nutrition.

(2) The hormone suppresses estrus, ovulation and the production of pituitary LH, which originally stimulated corpus luteum formation.

(3) Stimulates the mammary glands.

(4) Maintains the foetus by forming placenta and inhibits uterine muscles to contract during pregnancy.

The chief excretory product of progesterone is *pregnanediol* which is present in the urine of pregnant women. Progesterone is metabolised to pregnanediol in liver by a series of reductive steps.

Relaxin: In addition to progesterone, corpus luteum of some mammals like guinea pig, mouse, rabbit, dog and cow, reduce a second hormone termed *relaxin.* It relaxes pubic symphysis and thus facilitates birth.

Placenta and its Application

It secretes a hormone *Chorionic Gonadotropin* (CG) which is a product of the very early placenta (foetal part). Scientists have known for some time that Human Chorionic Gonadotropin exerts a protective influence on the unborn child, because deficiency or absence of the hormone in the early stages of pregnancy has been associated with abortion. This hormone has a major promoting influence on the production of progesterone from the ovary in the first six weeks. It is well known that progesterone is the gestational hormone which is essential for the continuation of pregnancy.

Function of Thymus

Thymus is partly an endocrine gland and partly a lymphoid structure. Located in the upper part of the thorax, the thymus is quite large in the young animal, but as the animal grows, the gland atrophies. By the time adulthood is reached, it is no more than a pair of small lobes lying on either side of trachea. Each lobe has numerous lobules and every lobule consists of a peripheral cortex densely packed with lymphocytes and a central medulla in which lymphocytes are less numerous.

The medulla also contains reticular cells and the unique

"Hassall's Corpuscle" which are spherical structures composed of concentric layers of spindle-shaped cells whose function is unknown.

Functions

(i) The thymus normally secretes a polypeptide hormone—thymosin or thymin which depresses neuro-muscular transmission (by regulating the amount of Arch at the neuro-muscular junction of skeletal muscles). An excess production of thymin by the abnormal thymus results in a disease myasthenia gravis in which the skeletal muscles are weak and there is a derangement of neuromuscular transmission.

(ii) The thymus in the infant mammal plays a major role in setting up the lymphocyte-producing machinery of the lymph nodes, thus providing the basis for the development of antibodies, in this way it keeps the owner immune from certain diseases.

Significance of the Pineal Body

The Pineal gland is a small, pea-sized structure lying deep in the groove between the cerebellum and the cerebral hemispheres in rabbit. Histologically, it consists of Parenchyme cells and Neuroglia cells.

In lower vertebrates, it produces a hormone called melatonin. When injected into frogs, it causes a marked lightening of the skin. Its action on the melanophores is thus just the reverse of that of MSH. In the Laboratory rat, melatonin inhibits the gonads both as gamete producers and as endocrine glands. As for humans, it atrophies at the age of 7 years and its role is uncertain.

Endocrine Glands in Insects and their Application

It has been established that the moulting and metamorphosis in insects is controlled by hormones. The insects have a well-developed compliment of endocrine organs consisting primarily of (1) the brain (protocerebrum) (2) the prothoracic glands and (3) the corpora allata.

The glandular cells of all three organs show secretory activity prior to each moult, and the three types of secretion are essential for the normal course of larval moults.

1. *The Brain (protocerebrum):* There are four groups of *neurosecretory cells* in the brain. Of these, two groups lie near the midline and one group on each side. The neurosecretory cells secrete a hormone which serves to activate the prothoracic gland as a result of which the moulting is initiated. The secretion of the brain, therefore, cannot act directly but only through the activation of the prothoracic gland.
2. *The Prothoracic Gland:* It is an irregular branching mass of glandular cells located in the thorax. The prothoracic gland lies in close association with the tracheal tubes. The hormone produced by the prothoracic gland is called the growth and moulting hormone or ecdysone. It has following functions:
   (i) It initiates the process of moulting in the epidermis as a result of which the epidermal cells undergo growth and proliferation.
   (ii) It is responsible for the shedding of the old cuticle and the production of the new one.

Butenandt and Karlson (1951) prepared ecdysone in a chemically pure form silkworm pupae. 25 mg of the pure substance was isolated from five hundred kilograms of silkworm pupae.

It was found that 0.0057 mg. of ecdysone thus isolated was enough to cause the pupation of a fly larva. Karlson and hoffmeister (1963) carried out the chemical analysis of the ecdysone and gave the empirical formula as $C_{27} H_{41} O_6$. Chemically, it is a steroid. Its structural formula is as follows :

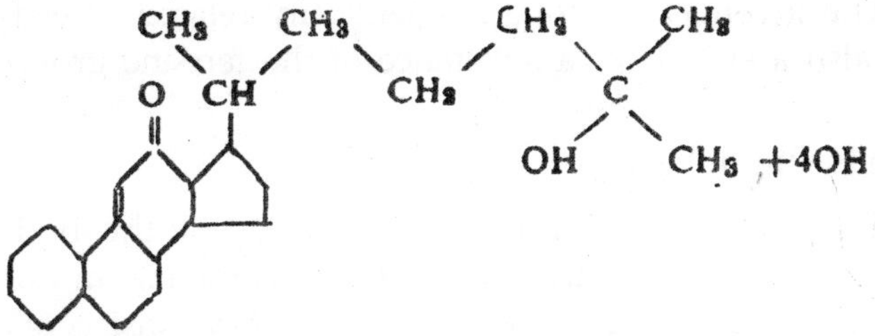

So far two ecdysones are now recognised: α -ectdysone and β -ecdysone.

3. *The Corpora Allata:* In most insects there are found two pairs of bodies connected by nerve strands to the protocerebrum and lying alongside to the dorsal aorta. First of these bodies on each side is the corpora cardiaca which is of the nature of nerve ganglia. Located more posteriorly is the corpora allata which consists of secretory cells.

The secretion of the corpora allata controls the nature of the change that is going to take place at the time of moulting. The hormones secreted by the protocerebral neurosecretory cells and the thoracic gland induce an insect to moult but they do not determine whether it will be a larval moult, the pupal moult, or the imaginal moult. On the other hand, the secretion of the copora allata controls the qualitative changes in the body of the insect during metamorphosis.

The presence of the corpora allata is necessary to prevent metamorphosis and to keep the insect in the larval state. Therefore, the secretion of the corpora allata has been called the juvenile hormone.

At every larval moult the cells of the corpora allata show signs of secretory activity but no such activity is observed during the pupal or imaginal moult. It is inferred, therefore, that at the time metamorphosis takes place, the corpora allata do not produce their hormone or at least are less active. It is actually the absence of the juvenile hormone that is the necessary condition for metamorphosis.

The juvenile hormone is chemically related to ecdysone and also a steroid or a substance of the terpene group.

Glands Physiology

The various physiological activities of the body are controlled by two systems: *nervous* and *endocrine*. In humans, as in all other mammals, the *endocrine system* consists of about ten different glands located in particular parts of the body. These glands have no ducts and liberate their secretions, known as the *hormones,* directly into the bloodstream. For this reason the endocrine glands are also referred to as the *ductless glands.*

Definition: A *hormone* may be defined as a *specific product (organic substance) of an endocrine gland secreted into the blood which carries it to some part of the body where it produces a definite physiological effect.* This effect may be either *excitatory* or *inhibitory* in its action. Hormones are not only the *"chemical messengers"* that pass from the glands via the bloodstream to the target organ or some process ; but they are important regulating substances that control virtually every aspect of the metabolism in living cells. There are also chemical coordinators other than hormones, such as CO_2 and other metabolites. Such substances are often called *Parahormones.*

Properties of Hormones

Hormones possess the following characteristics, which distinguish them from other chemicals:

(a) Hormones are produced by a specialised gland and are secreted directly into the bloodstream.

(b) Hormones exert their effects not where they are locally produced, but on other tissues of the body.

(c) Hormones act upon their target cells by regulating the rates of specific metabolic reactions.

(d) Hormones are required in very small amounts and their effect may be excitatory or inhibitory depending upon their concentration and the physiological state of the responding tissue.

(e) As against enzymes which are proteins, hormones have a varied chemical nature, including amino acid derivatives, polypeptides, steroids, etc.

Hormones Used as Chemicals

All the hormones are organic substances of varying structural complexity. Chemically, they may belong to any of the following categories:

(a) *Amino acid derivative: e.g.,* thyroxine.

(b) *Catecholamine: e.g.,* adrenaline, noradrenaline.

(c) *Polypeptide* or *protein: e.g.,* oxytocin, vasopressin, prolactin, adrenocorticotropic hormone, somatotropin (growth hormone), insulin, glucagon, secretin, relaxin.

(d) *Glycoprotein: e.g.,* thyrotropin, follicle-stimulating hormone, luteinising hormone.

(e) *Steroid: e.g.,* cortisone, aldosterone, testosterone, estrogen, progesterone.

Classifications of Hormones

On the basis of their control and influence over physiological processes, the hormones may be divided into five categories:

(a) Hormones concerned with the metabolism, e.g., insulin,

(b) Hormones of digestion, e.g., gastrin, secretin,

(c) Hormones that regulate the growth and development, e.g., growth hormone or somatotropin ;

(d) Hormones that control reproduction, *e.g.,* gonadotropic and sex hormones; and

(e) Hormones that control other endocrine glands, *e.g.,* thyrotropin.

Some hormones are called *local hormones;* for example, acetylcholine, secretin and cholecystokinin. These hormones have specific local effects. Other hormones are the so-called *general hormones,* which are secreted by specific endocrine glands. For example, epinephrine and norepinephrine, etc. Hormone concentrations in the blood are measured by *radioimmunoassay.*

Mechanism of Hormone Action: The following mechanisms have been suggested:

The cAMP Mechanism

A number of peptide hormones perform their action through cyclic AMP (3'-5'-adenosine mono-phosphate) and *adenyl cyclase.*

The Nobel Prize-winning work of E.W. Sutherland demonstrated that cyclic AMP, called as *second messenger,* mediates the action of many hormones and consequently regulates cellular activity in all forms of life. Cyclic AMP is produced by the catalytic action of *adenyl cyclase* on ATP.

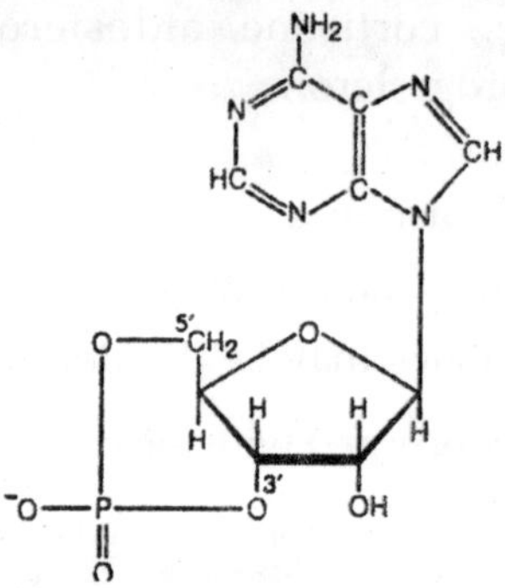

The structure of cyclic AMP.

$$\text{ATP} \xrightarrow{\text{adenyl cyclase}} \text{cAMP} + 2\ \text{Pi}$$

Once the hormone-activated reactions have been completed, cAMP is hydrolysed in the presence of Mg^{2+} to 5' AMP by an enzyme *phosphodiesterase.* All cyclic AMP-dependent reactions are turned off, because 5 AMP is ineffective on these targets.

Five components are involved in the mechanism by which ligands bring about changes in the intracellular concentration of cyclic AMP.

These are: a catalytic unit (adenyl cyclase), which catalyses the conversion of ATP to cyclic AMP; stimulatory and inhibitory receptors; and stimulatory and inhibitory G proteins (G_s & Gi that link the receptor to the catalytic unit, like the receptors *(receptor proteins)*, adenyl cyclase is a transmembrane protein and eight isoforms of this enzyme have been described.

Hormones affect target cells by binding to specific receptor proteins on the outer surface of the cell membrane. After the receptor binds its hormone, there is stimulation of adenyl cyclase (AC) associated with the inner surface of the membrane. Coupling of the function of the receptor to the adenyl cyclase is accomplished by another protein that is also bound to the inner surface of the membrane.

This protein, known as the *G-protein* strongly binds the nucleotide GTP when the receptor has bound hormone. With GTP bound, the a subunit of G-protein interacts with and stimulates or inhibits the activity of adenyl cyclase. When the appropriate ligand binds to a stimulatory receptor, a Gs α subunit activates the adenyl cyclase, and when it binds to the inhibitory receptor, a Giα subunit inhibits adenyl cyclase. The G-protein also has a GTPase activity.

Since adenyl cyclase is stimulated when a G_s-protein binds GTP, but not GDP, the GTPase activity tends to stop the stimulation of adenyl cyclase.

It appears that one hormone-receptor complex can activate several G-proteins to bind GTP and thus stimulate or inhibit several adenyl cyclases.

Two bacterial toxins (cholera toxin and pertussis toxin) have important effects on adenyl cyclase that are mediated by G-proteins. The drug *forskolin* stimulates adenyl cyclase activity by a direct action on the enzyme.

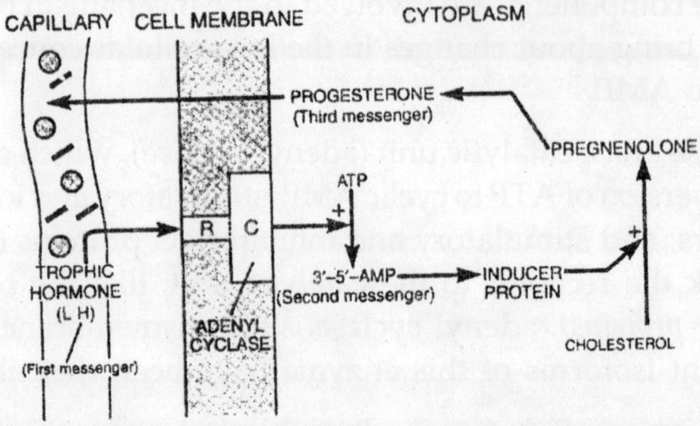

Mechanism of trophic hormone action on the cell to stimulate the production of cyclic-AMP, which then acts as a 'second messenger' to stimulate steroid secretion.

One interesting example of hormone-enzyme interaction is found in the liver cells where glucagon and epinephrine activate adenyl cyclase on the membrane of liver cell resulting in increased intracellular cAMP level. The increased cAMP concentration activates protein kinase which then converts by phosphorylation non-active enzyme phosphorylase *b* kinase into active phosphorylase *b* kinase. The phosphorylase *b* kinase is in turn phosphorylated to phosphorylase *a* which converts glycogen to glucose-1-phosphate.

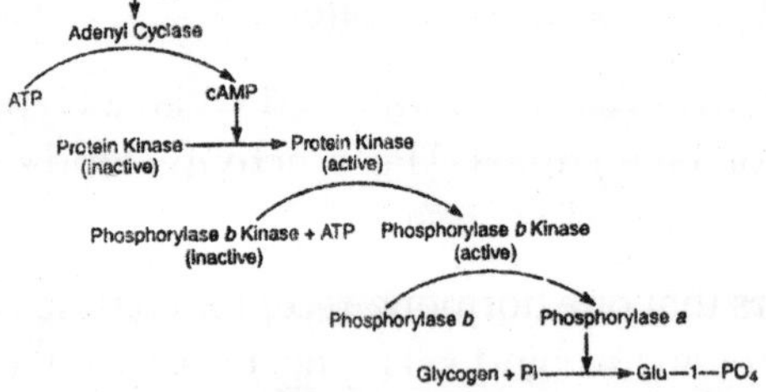

Activation of liver phosphorylase by hormones.

Other enzymatic processes that are known to be directly affected by the cyclic-AMP are glycogen synthetase, protein kinases, and phosphofructokinase, etc. Hormones which mediate their action through cyclic-AMP are polypeptides or amino acid derivatives including adrenaline, noradrenaline, glucagon, adrenocorticotropic hormone, thyroid-stimulating hormone, melanocyte-stimulating hormone, luteinising hormone, vasopressin, thyroxine, and parathyroid hormone.

(ii) Membrane Permeability Change

Some hormones might increase or decrease cellular activity by affecting the permeability of the cell membranes. All the neurotransmitter substances (local hormones) combine with receptors in the post-synaptic membrane. This causes a conformational change in the protein structure of the receptor, usually opening or closing a channel for one or more ions like Na, K or Ca, etc. Then, it is the altered movement of these ions through the channels that causes the subsequent effects on the post-synaptic cells. Some general hormones like epinephrine and non-epinephrine have especially powerful effects in opening or closing membrane ion channels for Na, K, or both, thus changing the membrane potentials of specific smooth muscle cells and causing excitation or inhibition.

(iii) Hormone-Gene Interaction

It is considered that some hormones act by controlling the gene activity in the target cell so that the DNA, through the genetic code mechanism, induces protein or enzyme synthesis that may produce physiological responses. *A steroid hormone* molecule released in the bloodstream combines with a receptor molecule present in the cytoplasm of the target cell (intracellular receptor).. The receptor molecule for the steroid hormone consists of two subunits and each subunit can bind to a single hormone molecule.

The receptor and the hormone molecule together form the hormone-receptor complex. This complex enters the nucleus and attaches itself to the target cell chromatin. The binding site occurs on a specific portion within the non-histone protein known as the AP_3 fraction. Thus, on entering the nucleus the hormone-receptor complex binds to AP_3 fraction with the second subunit attached to it. This combination determines what genes in the DNA molecule are going to be activated. The subunits then dissociate and the first subunit binds to the DNA enabling a molecule of RNA polymerase to occupy an initiation site at the DNA molecule. A segment of DNA is then transcribed producing a strand of messenger RNA that serves as a template for the synthesis of a protein. The latter in turn performs the necessary function of the hormone.

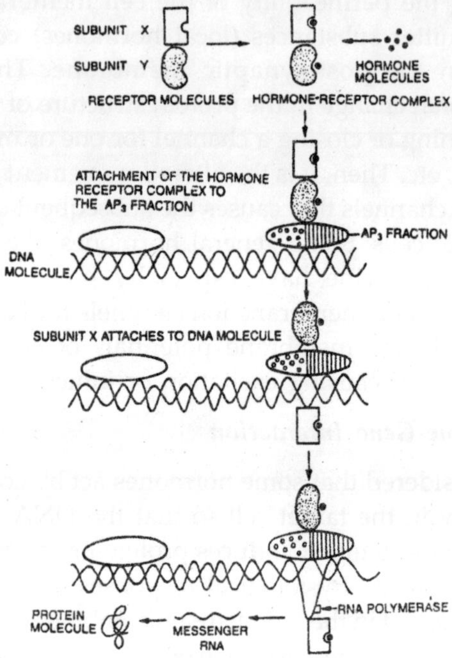

Mode of hormone action through protein synthesis.

As an example of hormone action of this type we may consider the activities of the *thyroid hormones,* which play an important role in temperature regulation in homeothermic vertebrates. Administration of small amounts of thyroid hormone to a mammal increases its metabolic rate. It is typical of this and other effects of thyroid hormones that the response to the treatment is delayed for several hours. This indicates that the calorigenic effect is secondary to some other effects of the hormones.

A study of the time course of events after the injection of a small, physiological dose of tri-iodothyronine into a thyroidectomised rat showed that increase in respiratory rate lagged behind an increase in protein synthesis in ribosomes and protein accumulation in mitochondria, which again lagged behind the synthesis of mRNA. It was, moreover, shown that treatment with actinomycin D, which is believed to suppress the formation of mRNA, completely prevented the calorigenic effect of the thyroid hormones. Treatments with other substances known to interfere with RNA or protein synthesis were also found to suppress or strongly reduce the calorigenic as well as the growth-stimulating effects of the thyroid hormones. The locus of action of these hormones thus appears to be at the level of the genes that produce the mRNA.

Although there are a number of instances wherein the ultimate hormonal effects turn out to be exerted on the level of synthesis of proteins, this mode of action does not appear to be universal.

Regulation of Hormone Secretion

The secretion of hormones by the endocrine glands is controlled by a complex of *chemical* and *nervous factors.* The secretory activity of most of the endocrine glands seems to be regulated by certain substances carried by the blood. In some cases these chemical factors are hormones themselves. In other cases they are relatively simple substances such as the sugars (*e.g.*, blood glucose stimulates the flow of *insulin* by the pancreas) and acids (*e.g.*, the secretion of *secretin* by the small intestine is stimulated by the acid chyme entering from the stomach). In a few glands (*e.g.*, the *adrenal medulla*) the secretory activities are regulated directly by the visceral nervous system.

Metabolic Clearance Growth

Two factors can increase or decrease the concentration of a hormone in the blood. One of these is the *rate of secretion* of the hormone. The other is the *rate of removal* of the hormone from the blood, which is called the *metabolic clearance rate.* It is calculated by the following formula:

$$\text{Metabolic Clearance Rate} = \frac{\text{Rate of disapperance of hormone from the plasma}}{\text{Concentration of hormone in each millilitre of plasma}}$$

Hormones are "cleared" from the plasma in several ways, including

(1) binding with the tissues,

(2) metabolic destruction by the tissues,

(3) excretion by the liver into the bile, and

(4) excretion by the kidneys into the urine.

Endocrine Glands in Invertebrates: Most of our knowledge of invertebrate endocrine glands is based on the investigations that have been recently made in crustaceans and insects. In general, it is difficult to homologise the invertebrate endocrine glands with those of the vertebrates. But many hormonal functions are analogous in the two groups and there is sufficient reason to believe that the hormones have an equally important role among the invertebrates.

The sows *gland* in the eye stalk was formerly regarded a gland of internal secretion among the decapod crustaceans. But it is now definitely known that a cluster of neurosecretory cells, called the *X-organ* located near the sinus gland, is actually the source of internal secretion. The axons of the neurosecretory cells of the X-organ convey the secretions to the sinus gland which functions as a storage and release centre for the hormones. The latter are released into the blood, which distributes them to the responsive tissues in the body. Several hormones are known to be definitely secreted by the X-organ. These are: (a) three *chromatotropic hormones* for the regulation of chromatophores or pigment cells, (b) *moult-inhibiting* hormone, and (c) the hormone for the control of eye pigment.

The process of metamorphosis in insects, like the process of moulting in crustaceans, is under the control of particular hormones produced by the neurosecretory cells and transported by the blood. Certain neurosecretory cells in the corpus brain of insects secrete the cardiacum brain hormone *(prothoracicotropic hormone)* which stimulates the so-called *prothoracic gland.* In *Periplaneta,* this endocrine gland is a X-shaped strand of endodermal cells in the ventral part of the prothorax.

The prothoracic gland then secretes *ecdysone,* a growth and differentiation hormone identified as a steroid. By initiating the complex process of growth and differentiation leading to moulting and the transformation from the larval to the pupal stage, *ecdysone* thereby initiates the post-embryonic growth and development of insects. Another hormone termed the *juvenile hormone,* secreted by a third endocrine gland, called the *corpus allatum,* lying near the brain, stimulates larval development but inhibits metamorphosis.

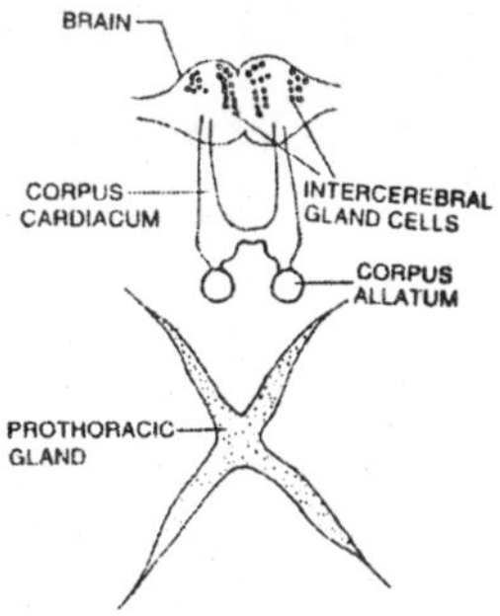

Endocrine glands in Periplaneta americana.

There is some evidence that the secondary sex characteristics among invertebrates are under the hormonal control. When parasitic castration by the parasite *sacculina* occurs in male crayfish, sexual changes towards femaleness are known to occur. According to their location in the body, the principal endocrine glands of the vertebrates belong to the following four categories:

1. *Cranial*: *e.g.,* the Pituitary gland.
2. *Pharyngeal: e.g.,* the Thyroid and the Parathyroid glands.

3. *Abdominal: e.g.,* the Islets of Langerhans, Adrenal, and Gastro-intestinal.
4. *Sexual: e.g.,* Testes, Ovaries and Placenta.

The Pituitary Gland or Hypophysis: *Location and Structure :* The pituitary gland is well protected, for it lies between the roof of the mouth and the floor of the brain in a depression of the sphenoid bone. It is a small rounded body about the size of a large pea weighing approximately 0.5 g in man but usually becomes much larger in a pregnant woman.

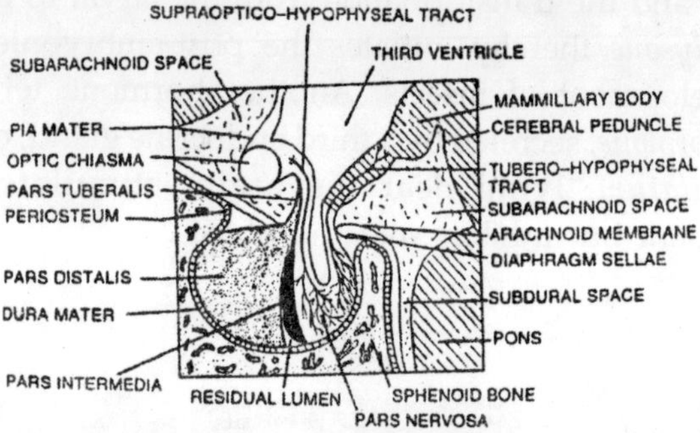

Section through a human pituitary and the adjacent structures at the base of the brain.

The adult pituitary gland is formed of two embryonic structures, one derived from the brain and the other from the epithelium of the mouth. Thus, it is a gland of double embryonic origin, consisting of two lobes, *i.e.,* the anterior and posterior lobes. The two lobes constituting the pituitary are essentially two different endocrine glands, as are the adrenal medulla and adrenal cortex, and the thyroid and parathyroid glands.

(a) *Anterior Lobe:* The upper or anterior lobe, now usually called the *adenohypophysis,* is derived from the epithelial tissue of the mouth. In the course of embryonic development the *neurohypophysis* moves downward, becoming posterior in position to the adenohypophysis which has developed upward to assume an anterior location relative to the neurohypophysis. The adenohypophysis in a mammal consists of the following parts:

(i) *Pars Tuberalis:* The pars tuberalis forms a collar of cells 25 to 60 μ thick around the neural stalk. It is thickest anterior to the stalk and frequently incomplete on the posterior aspect. The cells are arranged in short cords or globular clusters. They are polyhedral, with centrally placed nuclei and the cytoplasm contains fine granules, and is weakly basophilic. Nests of squamous cells are often found around the pars tuberalis. They are important because they may give rise to tumours (craniopharyngiomas).

(ii) *Pars Intermedia (middle or intermediate lobe):* In most species this portion of the adenohypophysis is quite distinct. However, in others *(e.g.,* the whale) and in birds, it is lacking. In man, the pars intermedia is rudimentary and its cells often surround colloid-filled cysts, and merge imperceptibly with those of the pars distalis. They are either *chromophobes* or *basophils.* The cells lining the follicles may be ciliated. A unique feature in humans is the invasion of the parsnervosa by basophil cells of the pars intermedia.

(iii) *Pars Distalis:* It forms about 75 per cent of the hypophysis. The cells in the pars distalis are arranged mainly in cords between which are large-bore capillaries. On the basis of the staining properties, three types of cells are distinguished in it, *acidophils, basophils,* and *chromophobes.* These are said to constitute about 40, 10, and 50 per cent of the anterior lobe (pars distalis) cells respectively.

(b) *Posterior Lobe:* The posterior lobe or the *neurohypophysis* has two subdivisions, *pars nervosa* or *infundibular process,* and *infundibulum* (median eminence and infundibular stem). While developing from the floor of the diencephalon behind the optic chiasma, the *infundibulum* and the *pars nervosa* at first contain the continuation of the cavity of the third ventricle.

This cavity is obliterated during the development except for cystic remnants lined by the ependymal cells. Following this obliteration, the hilar or central portion of the pars nervosa is formed by a densely packed bundle of non-myelinated fibres known as the *hypothalamohypophysial tract*. In man, however, the *median eminence* is not prominently developed, and it does not have the characteristic neurohypophysial structure seen in the other vertebrates, including the lower primates.

Hormones of Adenohypophysis

1. *Thyrotropin or Thyroid-Stimulating Hormone (TSH)*: This hormone has been purified to a considerable extent and is shown to be a *glycoprotein* with a molecular weight of less than 10,000. It controls various aspects of the thyroid gland function including the development and maintenance. Thyrotropin also stimulates iodine accumulation, the conversion of iodine into the thyroid hormones, and their release from the gland into the circulation. It is thought that certain nerve secretions from specific areas of the hypothalamus are transmitted by way of the pituitary stalk to regulate the secretion of thyrotropin. If there is an excess of secretion, there is an increase in the size of the thyroid gland. But excess of secretion is prevented by *thyroxine*.

2. *Adrenocorticotropic Hormone (ACTH):* The ACTH regulates the development and secretion of the adrenal cortex in the same way as TSH controls the thyroid gland. The ACTH molecule has been well characterised as a *polypeptide* of molecular weight 4500, containing 39 amino acids. The secretion of ACTH and adrenal cortical hormones are mutually affected by a negative

feedback mechanism. In addition, as in the case of TSH secretion, there is evidence of nerve secretions from certain hypothalamic areas of the brain which influence the secretion of ACTH. Such a mechanism suggests a basis for the release of ACTH by several different stimuli including the emotional stress, extreme temperatures, poisons and drugs, etc.

3. *Growth Hormone or Somatotropin (STH)*: This hormone is a protein. The growth hormone of humans has a molecular weight of 21500 and consists of 190 amino acids. The growth hormone stimulates growth of all the body cells including the skeletal and muscular tissues. Under-secretion of the growth hormone during the years of skeletal growth of an individual results in *dwarfism,* whereas an over-secretion of the growth hormone prior to adulthood result in *gigantism.* Over-secretion of the hormone after the attainment of full skeletal growth gives rise to the condition called *acromegaly (acros* = extremity, *megale* = large). It is manifested by a characteristic enlargement or overgrowth of the bones of the hands, feet, and in particular the jaws, cheeks and face as well as an increase in the overlying soft tissues.

4. *Gonadotropic Hormones (GTH)*: The three gonadotropic hormones secreted by the adenohypophysis are proteins. Of them, the FSH and LH are glycoproteins with a molecular weight of about 30,000 and with 15 to 18 per cent carbohydrate while the LTH is a polypeptide. Their secretion is apparently controlled by feedback mechanisms and by nerve secretions from the hypothalamus. These are:

   (a) *Follicle-Stimulating Hormone (FSH):* It is so-called because it stimulates the development and maturation of the ovarian follicles producing the egg cell and one of the female sex hormones, *estrogen.* In the male, the FSH induces spermatogenesis in the testes.

(b) *Luteinising Hormone (LH) or Interstitial Cell-Stimulating Hormone (ICSH):* In the female, LH is involved in the further development of the egg cell and its release by rupturing of the ovarian follicle. It is also responsible for stimulating the development of the *corpus luteum* that in turn secretes *progesterone.* The progesterone and estrogen are the two principal female sex hormones responsible for sexual drive and the development of the secondary sexual characters (*e.g.*, mammary glands, etc.). In the male, LH stimulates the secretion of the male sex hormone, *testosterone*, by specialised cells in the testes. Testosterone is primarily responsible for the sexual drive and development of the secondary sexual characteristics of the male (*e.g.*, hair on the body, deeper voice, etc.).

(c) *Lactogenic* (Luteotropic) *Hormone LTH* or *Prolactin* : Prolactin (PRL) is a polypeptide. It has been termed by *Nicoll* and *Bern* (1971) as "jack-of-all-trades", with more than eighty different hormonal activities ascribed to it. According to them, the many functions of prolactin belong to the five major categories: reproduction, osmoregulation, growth, integument, and a group of actions in which prolactin has a synergistic effect with the steroid hormones.

In mammals, this hormone induceds the milk secretion (lactogenesis) in the mammary gland of the mother shortly after giving birth. It is now known that prolactin also functions together with estrogen in promoting the development of the mammary glands. It also serves together with LH to maintain the corpus *luteum* (and therefore *progesterone* secretion) which is so vital for the maintenance of pregnancy. In the rat prolactin has luteotropic functions, *i.e.*, it promotes and maintains the secretion of progesterone. In bony fish, prolactin enhances production of a dilute urine and helps in osmoregulation by both gills and kidney.

5. *Intermedin or Melanocyte-Stimulating Hormone (MSH)*: Two distinct hormones known as the melanocyte-stimulating hormones have been purified from several vertebrates. They stimulate the distribution and concentration of the *melanin* pigment granules in the epidermal cells of the skin. In the frogs and tadpoles, MSH is responsible for the generalised blackening of the skin. Removal of the pituitary in these animals results in a concentration of the pigment near the centre of the cells, thus lending a lighter or silvery colour to the skin. However, the role of MSH in pigmentation in humans is not known.

6. *Diabetogenic Hormone:* It is also produced in the anterior lobe and affects the metabolism of carbohydrates.

7. *Ketogenic Hormone:* It causes an increased breakdown of fats to ketone bodies which appear in the blood and urine.

Hypothalamic Regulation of the Adenohypophysis: Many of the functions of the adenohypophysis depend on its connection with the hypothalamus. The hypothalamus produces three important substances, *gonadotropic release factor*, *corticotropic release factor* and *thyrotropic release factor* (or thyrotropin-releasing hormone, TRH), which govern the release of various tropic hormones on reaching the adenohypophysis by the hypophyseal portal system or by the general circulation. In return, these tropic hormones react back *(feedback)* and influence the functions of the nervous system. Thus, this plan involves a three stage hormonal system: releasing factors from the hypothalamus regulate tropic hormones from the adenohypophysis, which in turn regulate hormone secretions of the respective target glands, *e.g.*, gonads, thyroid, etc..

This mechanism of feedback, whereby a target organ can influence its own production positively or negatively by sending back information to an early stage in its synthetic pathway, is not clear in all instances. But it is thought that whenever a feedback occurs it is mediated through the nervous system and reaches the hypophysis by way of the hypothalamus. Apparently some pituitary hormones are not subject to feedback control and are secreted independently of the central nervous system. This general scheme shows that parts of the nervous system function as the specialised endocrine organs.

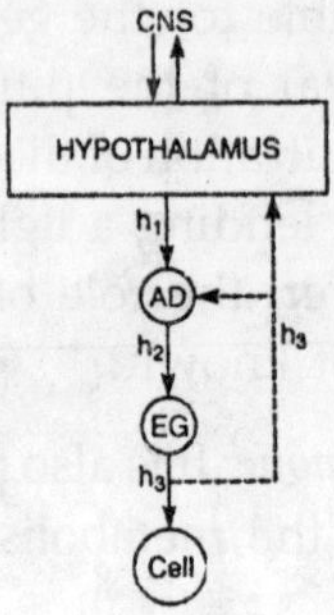

Hormones of Neurohypophysis

The neurohypophysis of most mammals, including man, releases *oxytocin* and *vasopressin.* Both vasopressin and oxytocin, are almost identical in structure with *isotocin, mesotocin* and *vasotocin,* the hormones released from this posterior lobe of the pituitary gland in the fish, frogs, and other non-mammalian vertebrates respectively.

However, the hormones of the neurohypophysis are secreted in the secretory nerve cells of the hypothalamus and migrate through the axons to the neurohypophysis where they are stored. It is generally believed that the nerve impulses control the synthesis of these hormones in the hypothalamus as well as their release from the neuro-hypophysis. Oxytocin and vasopressin are polypeptide hormones and they have seven amino acids in common. The similarity suggested that they arose through mutations in duplicate copies of a single gene. Vaso-pressin differs from oxytocin in having phenylalanine replacing isoleucine at position 3 in the ring, and arginine replacing leucine at position 8 in the side chain and it may be called 8-arginine-vasopressin. The structure is given in Fig.

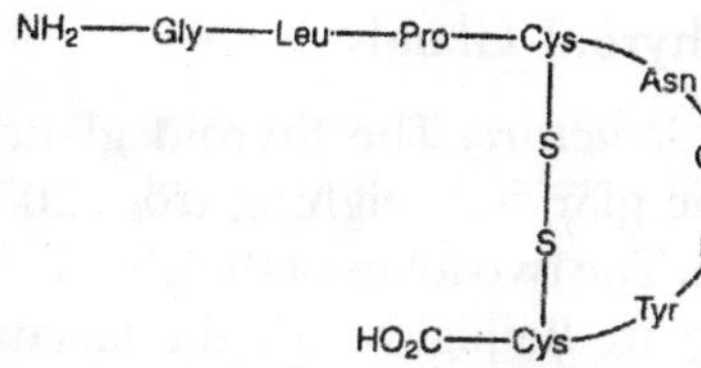

Oxytocin (Condensed Structure)

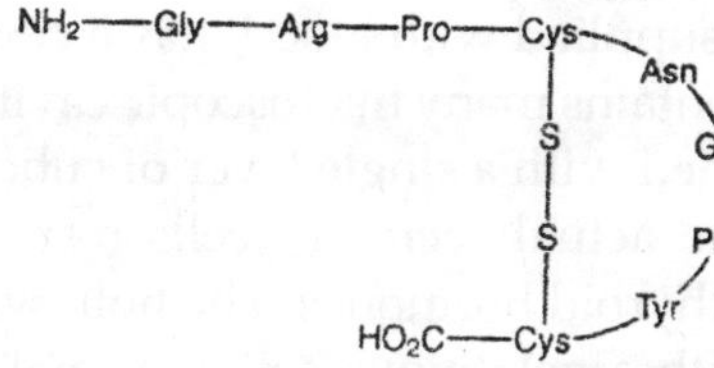

Vasopressin (Condensed Structure)

1 *Oxytocin:* It has a unique effect in producing the contraction of uterine muscles and uterus. It also promotes the ejection of milk from the ducts of the lactating breast. Several observations suggest that oxytocin may play a role in the reabsorption of sodium by the kidney, which may help to explain why the proportion of vasopressin and of oxytocin is the same in both sexes. Oxytocin is sometimes frequently injected during and after birth to stimulate contraction of the uterus and constriction of blood vessels, thus facilitating the birth process (parturition) and impeding the loss of blood.

2. *Vasopressin or Anti-diuretic Hormone (ADH):* It has an important role in regulating water reabsorption by the renal tubules. Apparently changes in the osmotic concentration of the blood and body fluids stimulate particularly sensitive neurons in the hypothalamus resulting in the release of the hormone from the neurohypophysis. An under secretion of vasopressin results in the disease *Diabetes insipidus*, in which a huge amount of urine is excreted. Vasopressin also causes the constriction of the musculature of the arterioles and capillaries, including the coronary and pulmonary vessels, thus causing some rise in blood pressure. For this reason it is termed *vasopressin.*

Significance of Thyroid Glands

Location and Structure: The thyroid gland is the largest of all the endocrine glands, weighing from 20 to 30 grams in adult human being. The two lobes of the gland, and the isthmus of tissue joining them, lie just below the larynx or voice box, just anterior to the upper portion of the trachea or windpipe. The gland is well supplied with blood, has a connective tissue framework, and contains many microscopic cavities or vesicles. The vesicles are lined with a single layer of cuboidal epithelial cells which are the actual secretory cells responsible for the production of the thyroid hormones. The hollow centre of each vesicle is filled with a gelatinous colloidal material made up mostly of a large complex glycoprotein called *thyroglobulin.*

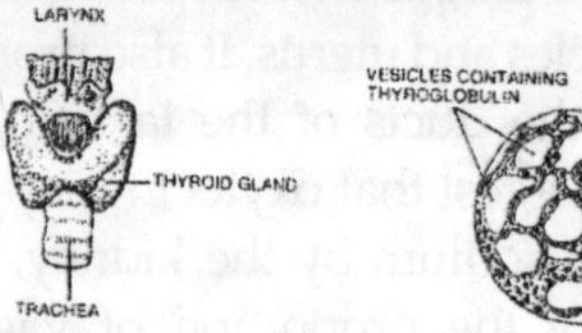

Gross and microscopic structure of the thyroid gland in man. Thyroid Hormones and their Physiological and Biochemical Effects

It is now known that the thyroid hormone consists of atleast four physiologically active iodinated organic compounds collectively called *iodothyronines* or thyroid hormones. The most important and abundant (about 93 per cent) of the four is called *thyroxine* or *tetraiodothyronine* (T_4) which until recently was believed to be the only physiologically active component or hormone secreted by the gland.

In human beings, thyroid glands also secrete *calcitonin,* which is a large polypeptide with a molecular weight of about 3400 and a chain of 32 amino acids. The so-called *parafollicular cells,* or C-cells, in the interstitial tissue between the follicles of the human thyroid gland which constitute only 0.1 per cent of the thyroid gland, secrete *calcitonin.* These cells are remnants of the ultimobranchial glands of lower animals. Thyroid secretion is controlled primarily by thyroid-stimulating hormone (TSH) secreted by the adenohypophysis.

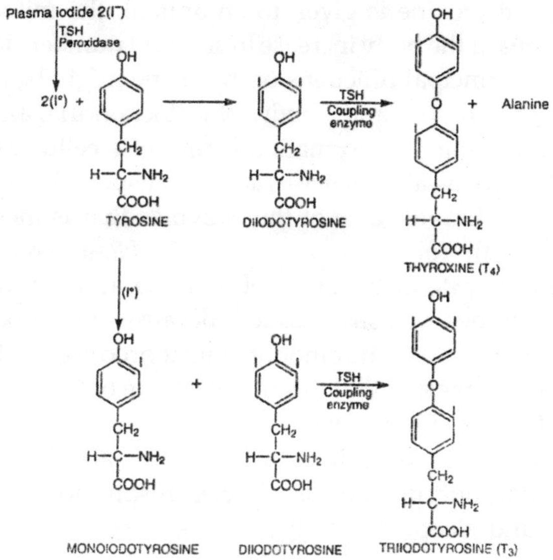

The synthesis of thyroid hormones (T_3 and T_4) from amino acid tyrosine. Before the iodide can be incorporated, it must be oxidised (I°) by a peroxidase system of enzymes, which is under the influence of TSH, thyroid stimulating hormone.

The structure of *thyroxine* (C_{15} H_{11} O_4I_4 N) is the iodinated derivative of the essential amino acid *L-tyrosine.* Its potency as a hormone is in some way associated with its four iodine atoms. If two of them are removed or, substituted by other atoms, potency is sharply reduced. If all the four iodine atoms are removed, potency disappears altogether. Being an amino acid, thyroxine is not digested in the gut, and it can, therefore, be taken by mouth.

Functions of the Thyroid Hormones

1. *Increased Matabolic Activity:* The thyroid hormones stimulate the rate of oxygen consumption or *oxidative metabolism* and resulting heat production (calorigenesis) of the various cells and tissues of the body. Thyroid hormone, in the homeotherms, stimulates many phases of *carbohydrate metabolism,* from the absorption of sugars to their utilisation in the tissues.

When thyroxine is given to an animal, the mitochondria in most cells of the body increase in size and number. Therefore, one of the principal functions of thyroxine might be simply to increase the number and activity of mitochondria, and they in turn increase the ATP formation to energise cellular function. Thyroid hormone also increases active transport of ions through the cell membranes. One of the enzymes that is increased in response to thyroid hormone is *Na, K-ATPase* which in turn increases the rate of transport of both Na and K through the cell membranes of some tissues. Because this process uses energy and increases the amount of heat produced in the body, it has been suggested that this might be one of the mechanisms by which thyroid hormone increases the metabolic rate of the body. Increased metabolism in the tissues affects blood flow and cardiac output and also affects respiration function of muscles and sleep.

2. *Regulatory Role:* Thyroid gland also has a second important function, namely a *regulatory role* (which it performs by its effects on the cellular metabolism) in general process of growth and development of all cells and tissues. These include the sexual development, maturation of bones and teeth, mental development, and energy metabolism, etc. The neo-natal requirement for thyroid hormones is, of course, well established, notably in human cretinism where thyroid deprivation results in reduced skeletal and somatic growth, retarded sexual development, mental deficiency, low basal metabolic rate, and increased risk of infection. However, the growth-promoting action of the thyroid hormones involves interaction with other hormones, especially somatotropin.

The effect of thyroid hormones on the nervous system is also quite clear. For example, if rats are made thyroid-deficient at birth, the cerebral cortex fails to develop normally. The behaviour of rats is also affected. Such rats are sluggish and perform poorly in learning experiments. However, the defects in the cerebral cortex can be corrected by thyroxine treatment, but only if the rats are treated within a critical period of about 15 days after birth. If the treatment is delayed, the defects are irrversible, as they are also in the humans.

3. *Effect on Other Endocrine Glands:* Increased thyroid secretion increases the rates of secretion of most other endocrine glands, but it also increases the needs of the tissues for hormones. For instance, increased thyroxine secretion increases the rate of glucose metabolism in the body and therefore a corresponding need for increased insulin secretion by the pancreas. Thyroid hormone also increases many metabolic activities related to bone formation and, as a consequence, increases the need for parathyroid hormone. Finally, thyroid hormone increases the rate at which adrenal glucocorticoides are inactivated by the liver. This leads to feedback increase in adrenocorticotropic hormone production by the adenohypophysis, and therefore, increased rate of secretion of glucocorticoid by the adrenal gland.
4. *Effect of Calcitonin on Blood Calcium:* Calcitonin, secreted by thyroid gland in human beings, tends to reduce the blood calcium ion concentration (an effect which is opposite to that of parathyroid hormone). In fish, amphibians, reptiles and birds, calcitonin is secreted by the *ultimobranchial* glands and plays an important role in controlling the blood calcium ion concentration when these animals change their habitat from freshwater to seawater (which contains more calcium).

However, it should be noted that calcitonin mechanism acts only weakly and only as a short-term regulator of Ca ion concentration because it is rapidly overridden by the much more powerful parathyroid control mechanism.

5. *Effect on Development:* Thyroid hormone also controls some developmental processes, at least in amphibia

and teleosts. In these vertebrates the hormone is specifically required in the *metamorphosis.* If thyroid function in an embryo or a young tadpole is in some way inhibited *(e.g.,* by excising the thyroid gland), then the animal remains a tadpole permanently. Conversely, if a young tadpole is given an excess of thyroxine, then the larva metamorphoses pre-maturely into a tiny froglet. Metamorphosis takes place normally when the thyroid gland has developed to maturity and produces appropriate amounts of the hormone. The developmental role of the thyroid hormone here resembles the role of the insect hormones.

Diseases of the Thyroid-Hypothyroidism: The phenomenon of under-secretion by the thyroid gland is called *hypothyroidism.* The occurrence of hypothyroidism (which may by due to a variety of factors including atrophy of the gland, or deficient dietary supply of iodine, etc.) in adults produces the condition called *myxedema,* a disease characterised by a low metabolic rate, increased plasma cholesterol, thickness and puffiness of the skin, coarseness and brittleness of the hair and finger nails, and a general physical and mental lethargy. There is usually a weight-gain, loss of hair, slower pulse rate, reduced blood pressure, and a decreased body temperature. The administration of adequate dose of the thyroid extracts quickly restores normal function to individuals with myxedema.

Cretinism: When *hypothyroidism* is present from the birth, the condition is called *cretinism.* It results from congenital lack of a thyroid gland *(congenital cretinism),* from failure of the thyroid gland to produce thyroid hormone, or from iodine deficiency in the diet *(endemic cretinism).* Cretinism is typified by retarded physical, sexual and mental development and low metabolic rate. Such children are stunted and of low intelligence. An early administration of the thyroid hormones by mouth or injection is necessary to prevent permanent effects of cretinism.

Goitre: Thyroxine deficiency may be accompanied by thyroid enlargement or *goitre.* Here the number of thyroid cells increases and more hormone may then be secreted by the enlarged gland. This may or may not suffice to compensate for an original hormone deficiency. If it does, the body functions will be normal but a *simple goitre* will be in evidence. If the deficiency is not compensated for, then cretinism in the young and myxedema in the adult may develop in spite of and in addition to the goitre. One of the causes of *simple goitre* is dietary deficiency of iodine and so the addition of iodine to certain foods *(e.g.,* table salt) and drinking water, or the administration of adequate amounts of thyroxine fronvan external source reduces its incidence.

Hyperthyroidism: The phenomenon of oversecretion by the thyroid gland is called hyperthyroidism or *thyrotoxicosis.* It is accompanied by physiological effects and symptoms which are the opposite of those observed in hypothyroidism. One of the most severe conditions of hyperthyroidism is called *Graves' disease* or *Exophthalmic goitre.*

It is characterised by an enlargement of the entire gland, an elevated basal metabolic rate, weight-loss, profuse perspiration, higher pulse rate, and a typical protrusion of the eyeballs called *exophthalmos.* The condition may be treated by surgical removal of a portiori of the thyroid gland, administration of radioactive iodine which accumulates in the thyroid destroying some of its tissues, or the use of certain thyroid-inhibiting drugs which interfere with the synthesis of thyroid hormones. Hyperthyroidism may also be caused by a tumour or growth of the thyroid gland.

Control of Thyroid Secretion

The various aspects of the thyroid gland function are controlled by *thyrotropin.* There are two main factors which influence the secretion of thyrotropin. They are as follows:

(a) The thyrotropin-releasing factor (TRF) of the hypothalamus, and

(b) The blood level of the thyroid hormones.

It is believed that the hypothalamus acts by secreting the TRF which is transmitted via the hypophysial portal system to stimulate the anterior pituitary cells. Thus, the secretion of thyroid hormones is apparently controlled by the thyroid-stimulating Hormone (also known as the thyrotropin or *thyrotropic hormone TTH)* secreted by the anterior lobe of the hypophysis. The thyrotropic hormone is produced in larger quantities as a result of low levels of thyroxine in the blood. The rise of thyroxine in the bloodstream in turn inhibits, by a negative feedback mechanism, the production of the thyrotropic hormone, thus decreasing stimulation of the thyroid. The regulation of the level of thyroid hormones once secreted into the bloodstream is affected in part by the liver through enzymatic mechanisms involving the removal of iodine and the amino group from the hormones.

The Parathyroid Glands Location and Structure: In human, the parathyroid glands consist of two pairs of small oval glands attached to the thyroid gland. These glands are absent in fishes, but are present in amphibians, reptiles, birds and mammals in either one or two pairs, depending on the species.

In adult human beings, the parathyroid gland consists of mainly *chief cells* and a small to moderate number of *oxyphil cells,* but oxyphil cells are absent in many animals and in young human beings. The chief cells are believed to secrete most of the parathyroid hormone. However, the function of oxyphil cells is not certain.

Parathyroid Secretion and its Function: The parathyroid hormone is a polypeptide with 84 amino acids and has a molecular weight of about 9500. Smaller ompounds with as few as 34 amino acids adjacent to the N terminus of the molecule have also been isolated from the parathyroid glands that exhibit full parathyroid hormone activity.

The parathyroid hormone tends to elevate the blood calcium level and decreases the phosphate level. The rise in Ca concentration is caused by two effects:

(1) an effect of parathyroid hormone in causing calcium and phosphate absorption from the bone *(osteolysis)* and

(2) a rapid effect of parathyroid hormone in decreasing the excretion of Ca by the kidneys. The decline in phosphate concentration is caused by a strong effect of parathyroid hormone on the kidneys in causing excessive renal phosphate excretion.

Parathyroid hormone also increases tubular reabsorption of Ca at the same time that it diminishes phosphate reabsorption. Parathyroid hormone greatly enhances both calcium and phosphate absorption from the intestine by increasing the formation in the kidneys of 1,25 dihydroxycholecalciferol from vitamin D. It should be mentioned that in the absence of vitamin D, the effect of parathyroid hormone in causing bone absorption is greatly reduced.

Hyperparathyroidism: It is produced by tumours on the gland and causes the calcium to be withdrawn from the bones, thus producing a high blood level of calcium. Under such conditions the bones may become soft and porous, or have only localised areas of calcification.

It should be noted that vitamin D is related in some unknown fashion to parathyroid function and the calcium and phosphate metabolism. For example, it can be shown that high doses of vitamin D can maintain normal blood calcium concentration in *Hypoparathyroidism*, because vitamin D promotes the rate of calcium absorption from the intestine. Thus, it can be noted that the bone cells withdraw calcium and phosphate from the blood for production of the bone, whereas parathormone withdraws these minerals from the bones. These two actions are antagonistic and tend to balance each other.

The Pancreas (The Islets of Langerhans)

Location and Structure: In human adults, the pancreas is an elongated gland weighing about three ounces. It lies behind the stomach close to the posterior abdominal wall, and functions both as a digestive gland and an endocrine gland. Besides the bulk of the glandular cells, which secrete a mixture of several digestive enzymes, the pancreas also contains scattered, rounded microscopic clusters or islets of tissue approximately 0.3 mm in diameter and about a million or so in number. These islets, discovered by *Langerhans* in 1869, are not connected with any ducts and perform an important endocrine function.

The *islets of Langerhans* in mammals consist of four known types of cells. The so-called *beta-cells* constitute majority (about 70-80 per cent) in the islets and secrete the hormone *insulin.* About 20 per cent of the islet cells are the *alpha-cells,* which are generally believed to secrete a second hormone called *glucagon.* However, the functions of the small percentage of remaining cells, called the C-cells, and D-cells are not known. D-cells may secrete the hormone *somatostatin.* In addition, at least one other type of cell, the PP cell, is present in small numbers in the islets and secretes a hormone called *pancreatic polypeptide.* Its function is uncertain.

Application of Insulin

The term insulin was given in 1916 and it was first extracted by Nobel Laureate *Banting* in 1922. Insulin is a proteinaceous hormone, consisting of 51 amino acid residues in the form of two linked peptide chains of 21 and 30 amino acids respectively with a molecular weight of 6000. The two chains are connected by two disulphide (-S-S-) linkages, while a third such linkage forms an intrachain bridge in the smaller chain. These disulphide bridges, which contribute to the tertiary structure of the molecule, are essential for its biological activity.

Insulin is rapidly digested by protein-digesting enzymes and must, therefore, be administered by injection and not by mouth to avoid its destruction in the digestive tract. The demonstration that insulin is actually present in the circulating blood (which is an essential condition for normal activity) depends in part upon the fact that plasma from normal animals can lower the blood-sugar levels of normal or diabetic ones.

The most convincing evidence, however, rests upon radio-immunoassay which makes possible the measurement of the circulating levels.

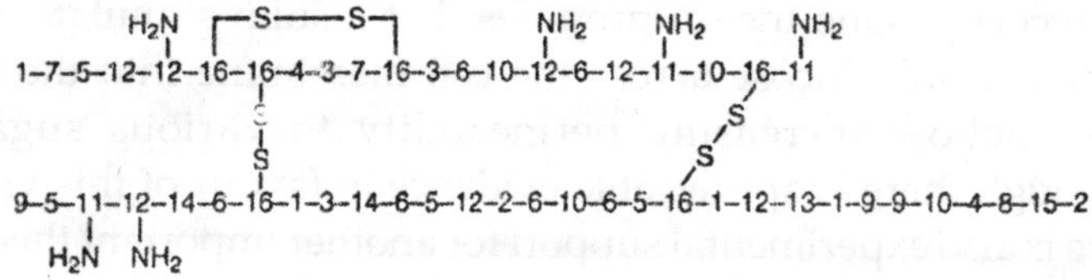

1. glycine	5. valine	9. phenylalanine	13. arginine
2. alanine	6. leucine	10. tyrosine	14. histidine
3. serine	7. isoleucine	11. aspartic acid	15. lysine
4. threonine	8. proline	12. glutamic acid	16. cysteine

Structure of the human insulin molecule.

The insulin is of great importance in the metabolism of fats and carbohydrates. Deficiency of insulin leads to excess sugar in the blood (300—1200 mg/dl) which is excreted in the urine. In this disease, known as *diabetes mellitus* meaning "a running through of sweet urine", much urine is excreted and the individual loses weight. When the blood glucose level rises to 300—500 mg/dl, which is common in people with untreated diabetes, 100 or more grams of glucose is lost into the urine each day.

As the diabetic can no more utilise the rich energy source of his high blood sugar, the person shifts from carbohydrate metabolism to fat metabolism for energy. As a result, there is an accumulation of incompletely oxidised ketone bodies from fat oxidation. These substances are toxic producing metabolic *acidosis,* and eventually lead to death of the patient.

Proper treatment of *diabetes mellitus* requires the administration of the right amount of insulin so that carbohydrate metabolism is balanced. Because insulin is a polypeptide and is not absorbed through the intestinal mucosa, it must be administered by intramuscular injection. Thus, insulin appears to be necessary for the storage of glucose or glycogen in the liver and for the oxidation of glucose.

But how does insulin stimulate the conversion of glucose to glycogen? One theory proposes that insulin stimulates the movement of glucose across the cell membrane into the cell by somehow increasing permeability to various sugars. Although there is appreciable evidence in favour of this view, there is also experimental support for another important theory. It proposes that insulin influences the early stage of glucose metabolism by possibly affecting hexokinase, which converts glucose to glucoses-phosphate (G-6-P) by phosphorylation utilising phosphate group from ATP.

Insulin has three sites of action. The first important field of action of the insulin is the *muscular tissue* where it acts to promote carbohydrate metabolism and the storage of glycogen. The glycogen can later be used for energy by the muscle.

The second important field of action of insulin is the *liver*. Here, in contrast to its action upon muscle, it does not affect the cell membrane, for the hepatic cells are freely permeable to glucose. In liver cells it favours glycogenesis and glycolysis, decreases hepatic gluconeogenesis, favours the uptake of fatty acids, and also increases lipogenesis.

When the quantity of glucose entering the liver cells is more than can be stored as glycogen or be used for local liver cell metabolism, insulin promotes the conversion of all this excess glucose into fatty acids. These fatty acids are subsequently packaged as triglycerides in very low density lipoproteins and transported in this form via blood to the adipose tissue and deposited as fat.

The third site of action of insulin is the *adipose tissue*, where it enhances the membrane transfer of glucose, which may then be stored as glycogen, or used immediately as the source of energy. It also enhances the uptake of fatty acids by adipose tissue, promotes lipogenesis, and decreases lipolysis. In all these three tissues insulin stimulates the uptake of amino acids and their incorporation into the tissue protein. It also facilitates the oxidation of glucose in most peripheral tissues. Insulin also affects protein metabolism and growth. It promotes, protein formation and prevents degradation of proteins.

Elevated glucose levels in the blood apparently serve as the stimulus for insulin secretion by the beta-cells of the islets. A lowering in blood glucose concentration is accompanied by a decrease or cessation in insulin secretion. The available quantity of insulin is the resultant of the rate of secretion and the rate of inactivation. A proteolytic enzyme, termed *insulinase,* is present in the liver, which appears to be the major site of insulin inactivation.

Hyperinsulinism and Insulin Shock: Increased insulin production resulting from adenoma of an islet of langerhans is called *hyperinsulinism.* When insulin causes the level of blood glucose to fall to low values (hypoglycemia caused by excess insulin), the metabolism of the central nervous system (which normally derives all its energy from glucose metabolism, and insulin is not necessary for this use of glucose) is depressed. Consequently, in patients with hyperinsulinism or in patients with diabetes who administer too much insulin to themselves, the syndrome called *insulin shock* may occur.

Functions of Glucagon

Glucagon or the *hyperglycemic factor* is another important hormone that is secreted by the *alpha-cells* of the islets of Langerhans. Glucagon (from the beef pancreas) is a relatively small protein molecule with a molecular weight of 3450, and consists of a single straight chain of 29 amino acid residues. A remarkable feature of the structure is that amino acid sequence of glucagon closely resembles that of secretin.

Whereas insulin in normally lacking in diabetes, glucagon is present in excessive amounts, and this excess of glucagon is probably as important a factor in diabetes as is insulin's lack. This hormone has the opposite effect of insulin, tending to increase the level of blood glucose by promoting the breakdown of liver glycogen (but not of muscle glycogen) to glucose *(glycogenolysis)*. The liver, therefore, appears to be the primary target organ for glucagon, which is believed to act by somehow enhancing the activity of the enzyme *(phosphorylase)* responsible for the breakdown of glycogen to glucose-1-phosphate. It possibly prevents enzyme inactivation or stimulates its activity. It also affects glucose metabolism by increased *gluconeogenesis* in the liver. Both these effects greatly enhance the availability of glucose to the other organs of the body. Thus glucagon provides an immediate defence against hypoglycemia and this defensive action seems to be one of its major functions.

Glucagon is also an important regulator of amino acid metabolism. The importance of amino acid stimulation of glucagon secretion is that glucagon then promotes rapid conversion of amino acids to glucose, thus making even more glucose available to the tissues.

The secretion of glucagon is regulated by a feedback mechanism similar to that which controls insulin secretion, but it works in the opposite direction. Low blood sugar stimulates the alpha-cells of the islets of Langerhans to secrete glucagon, whereas high blood sugar inhibits them.

Function of Somatostatin

Somatostatin or the *somatotropin-release inhibiting factor* has been isolated from sheep brain in the 1973 by *Brazeau* and his associates. It has recently been reported that somatostatin is not an exclusively hypothalamic peptide but may be synthesised in the pancreas by the *delta cells* of islets of Langerhans. *Dubois* (1975) has shown immunoreactive somatostatin in the endocrine cells of human foetal pancreas.

Besides inhibiting the release of somatotropin (growth hormone) somatostatin also acts on endocrine pancreas to inhibit the secretion or glucagon and insulin. It has been shown by several investigators that somatostatin reduces the concentration of insulin, glucagon and glucose in the blood of healthy individuals. Recent experiments with somatostatin strongly indicate that it is possible to alleviate the symptoms of diabetes by reducing the concentration of glucagon by somatostatin.

Adrenal Glands and their Hormones

Location and Structure: Each adrenal or suprarenal gland in man adrenal medulla weighs about 6 grams, measures 25 to 50 mm in length, and lies at the top of each kidney. The adrenal glands are somewhat pyramidal in shape and have a rich blood supply. They are innervated by the sympathetic nervous system, and consist of two parts that are embryologically, anatomically and functionally different. In the typical mammal, each gland consists of an external reddish brown *cortex* which surrounds an internal grayish *medulla*. The adrenal cortex has arisen from the same origin *(mesodermal)* as that of the sex glands, whereas the medulla has an ectodermal origin similar to that of the post-ganghonic fibre cells of the sympathetic nervous system.

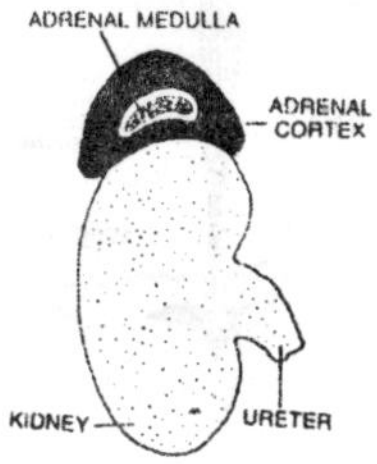

Showing the position and internal structure of the adrenal gland in man.

The cortex is made up of three zones, the cells of which differ in form and function. The outermost *zona glomerulosa* is formed by closely packed, low columnar cells and it secretes mineralocorticoids, hormones concerned with water and electrolyte mechanism and the most important of which are aldosterone and deoxycorticosterone. The secretion of these hormones is mainly regulated by negative feedback through plasma Na^+ and K^+ levels, but there is some evidence that ACTH may also be involved in their regulation. The thick, middle *zona fasciculata* is formed by cords of cells filled with lipid, the amount of which varies with the secretory activity of these cells.

The third or the innermost zone is the *zona reticularis.* The middle and inner zones secrete glucocorticoids, hormones that influence carbohydrate and protein metabolism. Cortisol is the most important glucocorticoid, and its secretion is regulated by ACTH.

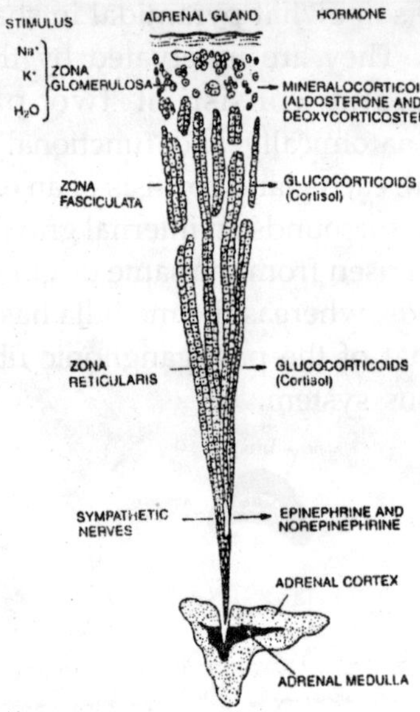

Showing three zones of adrenal cortex and hormones secreted by them.

Hormones of the Adrenal Cortex: The adrenal cortex is absolutely essential to life. When it is completely removed most animals will not live more than a week or so. The cortex has several hormones and it ranks next to the adenohypophysis in endocrine importance. Secretion of the cortical hormones is under the control of ACTH from the pituitary. All the *adrenocortical hormones* are steroid compounds and can be divided into three groups.

(i) *Glucocorticoids* such as *cortisone, hydrocortisone* (cortisol), and *corticosterone.*

(ii) *Mineralocorticoids,* such as *aldosterone* and *deoxycorticosterone.*

(iii) *Androgens,* sex hormones which have chiefly *androgenic* and to a minor extent *estrogenic* effects.

The adrenal cortex is rich in *ascorbic acid* (vitamin C) and *cholesterol,* the chief compound from which the steroid hormones are synthesised.

Synthesis of Corticosteroids

The adrenal steroids originate from cholesteral through pregnenolone or progesterone. The main changes are additions of hydroxyl groups. The addition of one hydroxyl group to *progesterone* at C_{21} forms 11 deoxycorticosterone, a mineralocorticoid. Further hydroxylation at C_{11} produces a glucocorticoid, *corticosterone.* If Q_{17} is also hydroxylated, the most potent naturally occurring glucocorticoid, *hydrocortisone* is formed. A rather more complex pathway is needed for the synthesis of *aldosterone.*

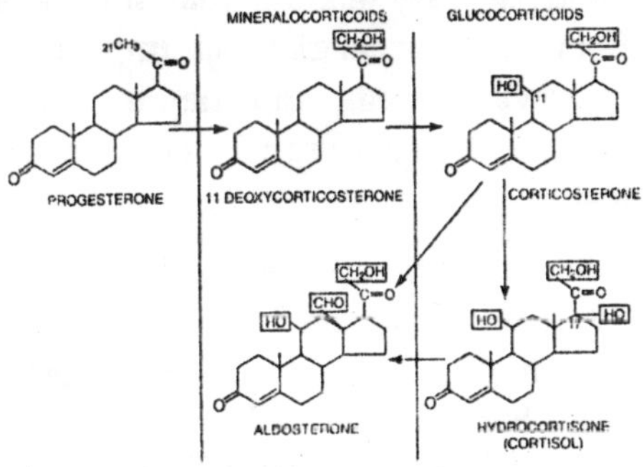

Origin and interrelationships of adrenocortical hormones.

Functions of the Adrenocortical Hormones

1. *Effect on Carbohydrate and Protein Metabolism:* The glucocorticoid hormones exert their influence primarily on carbohydrate metabolism. They increase the blood

glucose level at the expense of liver glycogen and enhance the conversion of protein (and fat) to carbohydrates. Cortisol also affects fat metabolism.

2. *Effect on Mineral, Salt, and Water Metabolism*: The mineral corticoids have a regulatory effect on the relative concentrations of mineral ions (especially sodium and potassium) in the body fluids, and therefore on the water contents of the tissues. The insufficiency of the mineral corticoids results in an increased excretion into the urine of sodium ions, chloride ions, and water, accompanied by a fall in blood sodium chloride, and bicarbonate. The administration of the mineral corticoid hormones has the opposite effects, resulting in an increase in blood sodium and chloride concentrations, more retention of water by the body, a decrease in potassium concentration, and a restoration of pH of the blood to its more normally alkaline level.

Aldosterone also promotes sodium retention in the salivary glands, sweat glands, and colon, and the excretion of *magnesium* in the urine. Thus *aldosterone* has almost the same effects on sweat glands and salivary glands that it has on the renal tubules. The effect on the sweat glands is important to conserve body salt in hot environments, and the effect on the salivary glands is necessary to conserve salt when excessive quantities of saliva are lost.

3. *Effect on Sexual Development:* Under certain circumstances the adrenal cortex appears to secrete steroids which are similar in their effects to male hormones. For example, adrenal cortical tumours in females may give rise to masculine characters, such as beard and deep voice, etc. Although the female sex hormones have also been detected in cortical extracts, it is believed that they are simply intermediates in the formation of other adrenocorticosteroids.

4. *Anti-allergy, Anti-inflammatory and other Effects:* Several

of the cortical hormones (*e.g., cortisol* and *cortisone*) have a marked influence in preventing the appearance of an inflammation. Some cortical hormones (*e.g., cortisol*) also counteract the symptoms of allergic or hypersensitive reactions. Accordingly, they have been used in the treatment of several kinds of acute inflammatory and allergic diseases especially of the eye. The two basic anti-inflammatory effects of *cortisol* are:

(1) it can block the early stages of the inflammation process before inflammation even begins, and

(2) if inflammation has already begun, it causes rapid resolution of the inflammation and increased rapidity of healing. They are also used to control the symptoms of particular arthritic conditions and rheumatic fever.

The steroid secretions of the adrenal cortex also exercise several other influences on the body functions including a role in decreasing the level of red and white blood cells and stimulating the secretions of the digestive tract.

Characteristics of Hypoadrenalism

Insufficiency of the cortical hormones results in *Addison's disease*. In this disease the adrenal cortex degenerates, resulting in a number of symptoms, including a characteristic bronzing of the skin, muscular weakness, low blood pressure, and digestive disturbance. The sufferers of this disease are cured by a proper administration of cortical hormones from some external source.

Significance of Hyperadrenalism

Hyperfunction or oversecretion by the adrenal cortex as a result of excessive secretion of ACTH, gives rise to *Cushing's disease*. The symptoms include high blood pressure, salt retention, swelling of the tissues with water, demineralisation of the bones, and loss of sexual function. Oversecretion may also arise from a tumour in the cortical cells and the resulting symptoms will depend on the kinds of hormones secreted by the tumour. If glucosteroids are predominantly produced, metabolism is affected. Treatment in Cushing's disease consists of removing an adrenal tumour if this is the cause, or decreasing the secretion of ACTH, if possible. If more concentrations of male-like hormones are secreted, the females begin to attain masculine characteristics and males show a greater accentuation of the male sexual characteristics.

Hormones of the Adrenal Medulla: The adrenal medulla in man consists of two distinct types of cells. These are *adrenaline-secreting* cells and the *noradrenaline-secreting* cells which are responsible for the production of the two hormones, *adrenaline* or *epinephrine* and *norepinephrine* or *noradrenaline* respectively. In human beings, these catecholamines of the medulla comprise about 80 per cent adrenaline and 20 per cent noradrenaline.

They are secreted in response to sympathetic stimulation. In turn, these hormones cause almost the same effects as direct stimulation of the sympathetic nerves in all parts of the body.

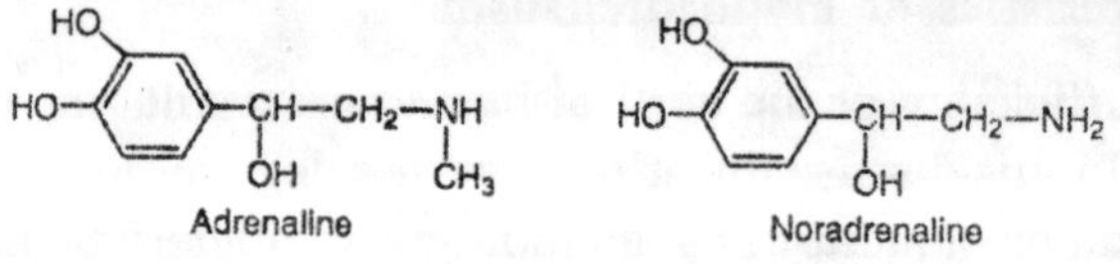

The structure of adrenal medulla hormones.

Adrenaline (Epinephrine): It is one of the simplest hormones known and can be produced synthetically. It was the first hormone to be isolated in crystalline form in 1901 and was first synthesised in 1904.

It functions as the so-called "emergency" hormone for the body because the general secretion of this hormone centres around emergency functions of the body, such as those of fear, fight, and rage, etc. During emotional states epinephrine is secreted in great amounts. It has a decided effect in elevating the glucose level in the bloodstream by somehow enhancing enzymatic conversion of liver glycogen to glucose. The rise in blood glucose is accompanied by an increase in oxygen consumption, body temperature and heat production. Adrenaline also tends to increase the flow of blood in skeletal muscles, heart and viscera by dilating the blood vessels. In this way it serves to direct the circulation of blood where necessary during exertion or increased activity.

It also increases blood pressure, heart beat, inhibits peristatic movements of the gut, stimulates the tiny muscles which attach to hair and feathers (appendages which are raised as a consequence producing "goose flesh" in man), dilates the pupils of the eye, increases muscular power and resistance to fatigue, promotes faster coagulation of blood, and generally promotes faster and sharper responses to external stimuli.

Together, most or all these effects are sometimes called the *alarm reaction*. If the action of adrenaline is somehow inhibited, a vertebrate cannot adjust rapidly to emergencies.

Noradrenaline (Norepinephrine): This hormone of the adrenal medulla has somewhat similar effects. It raises the blood pressure, but it is responsible for constriction instead of dilation of the blood vessels. Noradrenaline is a chemical transmitter at sympathetic nerve endings and in the central nervous system, but is also present in the adrenal medulla, both as a hormone and also as a precursor of adrenaline.

The following differences between the actions of epinephrine and norepinephrine must be noted.

1. Epinephrine, because of its greater effect in stimulating the beta receptors, has a greater effect on cardiac stimulation than does norepinephrine.
2. Epinephrine causes only weak constriction of the blood vessels in the muscles, in comparison with much stronger constriction caused by norepinephrine. This difference is of special importance because norepinephrine greatly increases the total peripheral

resistance and elevates arterial pressure, whereas epinephrine raises the arterial pressure to a lesser extent but increases the cardiac output considerably more because of its excitatory effect on the heart.

3. Their effects on tissue metabolism are also different. Epinephrine has 5 to 10 times as great a metabolic effect as norepinephrine. Epinephrine can increase the metabolic rate of the whole body often to as much as 100 per cent above normal, thus increasing the activity of the body. It also increases the rate of other metabolic activities such as glycogenolysis in the liver and muscle, and glucose release into the blood.

Gastro-intestinal Hormones

Several hormones are associated with the activities of the digestive tract. All of these hormones are produced in the mucosa lining and help regulate the volume and character of the secretions. Chemically, these hormones are polypeptides or polypeptide derivatives.

Gastrin: When food enters the stomach, *gastrin* (a hormone produced in the pyloric wall) stimulates the gastric glands to secrete gastric juice. Two closely related peptides (gastrin I and II) are present in the stomach of various mammalian species, including man, sheep, pig, and dog, etc. These are heptadecapeptide amides, gastrin II having a sulphate group attached by ester linkage to the single tyrosyl.

Secretin: The secretion of pancreatic juice, like that of the gastric juice, is regulated by both neural and hormonal factors. The entrance of acid chyme from the stomach into the duodenum causes the release of *secretin* from the cells of the duodenal wall into the blood circulation.

Secretin is a single-chain polypeptide of low molecular weight with 27 amino acid residues and bearing a strongly basic charge due to the arginine and histidine residues, and to the amidation of two of the glutamic acid ones. The arrival of *secretin* at the pancreas by way of the bloodstream stimulates the secretion of copious volumes of pancreatic juice low in digestive enzymes. *Secretin* also stimulates the flow of bile from the gall bladder and intestinal juice from the glands in the wall of the small intestine.

Pancreozymin: A second intestinal hormone called *pancreozymin,* stimulates the secretion of pancreatic juice rich in digestive enzyme activity.

Cholecystokinin: The entry from the stomach of fatty foods into the upper part of the duodenum promotes the release into the bloodstream of an intestinal hormone called *cholecystokinin.* It causes the gall bladder to contract discharging its contents into the cystic and common bile ducts.

Cholecystokinin/pancreozymin is a single-chain polypeptide with 33 amino acids and with a sulphated tyrosyl. As with gastrin and secretin, it has a number of actions additional to its primary ones upon the gall bladder and the secretion of pancreatic enzymes. It stimulates secretion of gastric acid and of the duodenal succus entericus, it inhibits gastric motility, and it increases duodenal motility.

Enterocrinin and Duocrinin: There is some evidence that the secretion of intestinal juice is under the control of two specific intestinal hormones called *enterocrinin* and *duocrinin,* which are secreted by the intestinal lining into the bloodstream.

Enterogastrone: The entry of lipids into the duodenum apparently causes the liberation by the intestinal wall of a hormone called enterogastrone which inhibits gastric juice secretion.

Gastric Inhibitory Polypeptide: GIP is a well characterised product of the small intestine, inhibiting both gastric mobility and gastric secretion. In has 43 amino acids, the sequence resembling with secretin and glucagon. Fifteen of the first 26 amino acids at the NH_2-terminal end are in the same position as in porcine glucagon, while 9 of the first 26 amino acids are in the same position as in secretin. It has been suggested that it is identical with enterogastrone. However, no deficiency diseases have been associated with the under-secretion of the gastro-intestinal hormones and some endocrinologists consider them of minor importance.

Gonads and their Characteristics

Location and structure: The specialised organs which eventually produce the reproductive cells (germ cells) are termed the *gonads.* In higher animals, the male gonad produces the sperm cells and is called the testes. The female gonad which produces the egg cells or ova is called the *ovary.* The mature gonads of both sexes are not only the centres for production of the germ cells but they also secrete the *sex hormones* which control the maturation and function of the reproductive system and certain other tissues of the body.

Testes and their Application

Each of the two testes in the adult male is an ovoid body about the size of a walnut. Both are located outside the body cavity within a skin covered pouch called the *scrotum.* The wall of each testis consists largely of connective tissue which is continuous with wall-like partitions dividing the interior of the testis into approximately 250 small compartments called *lobules.* Each lobule contains from one to three tiny convoluted seminiferous tubules within which the sperms are formed. The special cells located among the *seminiferous tubules* are collectively called the *interstitial tissue,* which performs the important function of producing and secreting the *male sex hormone.* The interstitial cells are not present during childhood but are found after puberty.

Role of Ovaries

Each of the two ovaries in the adult female is about the size and shape of large almond and is separately located in the lower or pelvic region of the abdominal cavity. Each ovary consists of a large number of *ovarian follicles.* The ovarian fallicles produce *the female sex hormones* while undergoing maturation.

Male Sex Hormones and Their Physiological Effects: The male sex hormones are usually referred to as androgens, including *testosterone, dihydrotestosterone* and *androstenedione.* The term "androgen" means any steroid hormone that has masculinising effects. The male sex hormones have not only been obtained from the testes but from the adrenal cortex, placenta, and the ovaries as well.

The principal male sex hormone is *testosterone.* It is a steroid, and is secreted by the interstitial cells of the testes of most mammals, including man. Much of the testosterone that becomes fixed to the tissues is converted within the cells to *dihydrotestosterone,* especially in certain target organs such as the prostate gland in the adult.

Some important target tissues do not have the reductase enzyme in their cells to convert testosterone into dihydrotestosterone. In these tissues, testosterone functions directly. For instance, this direct effect of testosterone is essential in the male fetus for the development of the epididymis, vas deferens, and seminal vesicles. The direct effect is also responsible for much of the effects of testosterone to promote spermatogenesis.

The production of testosterone is stimulated by such gonadotropins as the luteinising (or the interstitial cell-stimulating) hormone. The basic intracellular mechanism of action of testostrone is that it greatly stimulates production of proteins virtually everywhere in the body, although increasing in particular those proteins in target organs responsible for the development of secondary sexual characteristics.

Testosterone is responsible for the secondary male sex characteristics, such as the beard of the face, the distribution of hair on the body, and the character of the voice, etc.

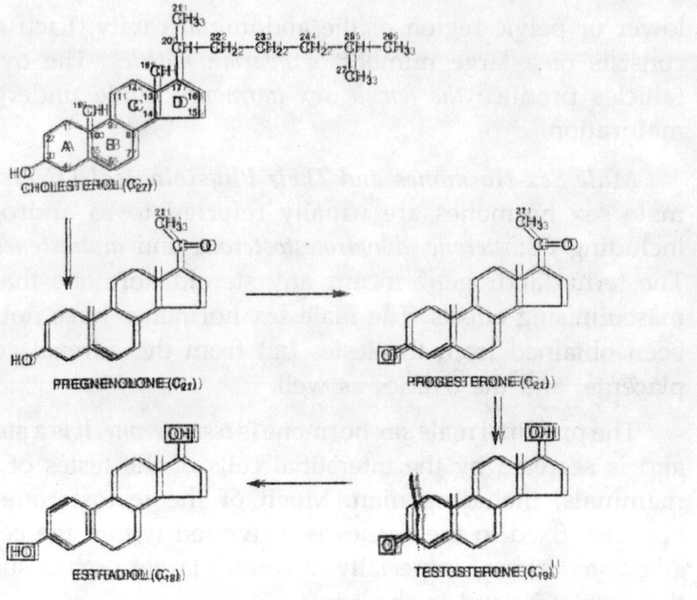

The structural relationship and possible order of synthesis of steroid hormones from cholesterol.

It is also concerned with the growth and development of the accessory sex characters, penis, seminal vesicle and prostate gland. When injected into a female, secondary male sex characteristics are often developed. Testosterone is also responsible for the sex urge and for sexual behaviour. Testes-removal or *castration* in the human male prior to the onset of puberty results in the failure of the secondary sexual characteristics to develop. After puberty, castration causes an attenuation of the sexual characteristics.

Female Sex Hormones and Their Physiological Effects: The ovary secretes two types of female sex hormones called *estrogens* and *progestins*. By far the most important of the estrogens is the hormone *estradiol*, and by far the most important progestin is *progesterone*.

Estrogen: Estrogen is a collective name for three different but closely related steroid hormones. These are β-*estradiol*, *estrone*, and *estriol*. The most common ovarian estrogen of many mammals is considered to be *β-estradiol*. The estrogenic potency of β-estradiol is 12 times higher than estrone and 80 times that of estriol. Starting shortly before puberty and continuing throughout the sexually mature life of the female, the secretion of estrogen is periodically stimulated by the follicle-stimulating hormone (FSH) of the adenohypophysis.

Another important source of estrogen in humans and many other mammals, second to the developing follicles of the ovary, is the *placenta*. Estrogen is also present in the adrenal cortex. However, it has not only been isolated from various invertebrates, but also from plants. About 30 families of higher plants have been shown to contain one or more forms of estrogen. Whether they perform a significant role in plants or are simply non-functional by-products of steroid metabolism is not known.

Estrogen is responsible for the mature growth, development and maintenance of the female reproductive tract, the secondary sex organs (breasts, vagina, uterus) and other secondary sexual characteristics. It also contributes to the sexual desire of the female. Estrogen promotes repair of the uterine lining following menstruation in preparation for possible implantation of a fertilized egg. Although the hormone helps to control the development and functioning of the mammary gland, it appears to inhibit lactation by inhibiting the secretion of the lactogenic hormone by the pituitary gland. In general, estrogen tends to decrease the secretion of the gonadotropic hormones of the pituitary.

After the monopaure almost no estrogens are secreted by the ovaires. This estrogen deficiency leads to diminished osteoblastic activity in the bones, decreased bone matrix, and decreased deposition of bone calcium and phosphate. In some old women, this effect is extremely severe, and the resulting condition is called *osteoporosis* of bones.

Estrogens also affect protein and fat deposition, hair distribution, skin, and electrolyte balance.

Progesterone: Progesterone is the second principal female sex hormone. It is also a steroid somewhat related in chemical structure to estrogen. It is secreted during the later half of the menstrual cycle in human females chiefly by the temporary endocrine tissue, the *corpus luteum,* which rapidly develops in the ruptured follicle shortly after the release of the egg cell.

Progesterone further promotes the secretory changes in the uterine endometrium already initiated by the estrogen during the first half of the menstrual cycle, in preparation for implantation of the fertilized ovum. At this particular phase of the female menstrual cycle, it takes over the general functions of estrogen (when the latter is being secreted at a minimum level) in maintaining the mature female reproductive tract and secondary sexual characteristics. Thus, together with estrogen, progesterone is responsible for the sequence of changes of the repeating menstrual cycles that occur in the female reproductive tract starting at puberty and extending through the reproductive life of the female.

Progesterone also affects fallopian tubes and breasts. It promotes development of the lobules and alveoli of the breasts, causing the alveolar cells to proliferate, enlarge and become secretory in nature. However, it does not cause the alveoli actually to secrete the milk which is secreted under the stimulus of prolactin.

Like some of the adrenal cortical hormones, progesterone also seems to promote retention of sodium salts and water (electrolyte balance), presumably by affecting kidney function. Progesterone is also found in the mammalian placenta, the adrenal cortex, and the testes of some mammals (*e.g.*, the pig). The occurrence of the hormone in the adrenal cortex and testes is probably due to its role as an intermediate in the synthesis of other steroid hormones.

Relaxin: Another female sex hormone called *relaxin* is produced by the corpus luteum of the ovary and by the placenta. It is a polypeptide, having a molecular weight of about 9000. It specifically causes the relaxation of pelvic ligaments in child birth. It has also been claimed that relaxin softens the cervix of the pregnant woman at the time of delivery.

Role of Placenta

The human placenta, which has been extensively studied, has been described as an 'incomplete endocrine gland' because it is deficient in many of the essential enzymes necessary for complete synthesis of steroid hormones from acetate. The importance of the human foetus in the production of placental hormones has been recognised by the use of the term 'foeto-placental unit' or the 'foeto-placento-maternal unit'. The latter term appears more appropriate as both the maternal and foetal liver and adrenal glands supply preformed precursors to the placenta, which converts them into active hormones. The placenta of some mammals secretes the following protein and steroid hormones:

Progesterone: Human placenta probably converts cholesterol or pregnenolone from the maternal blood to progesterone which is secreted by the placenta in large amounts and is important in maintaining pregnancy. Besides inducing widespread adaptations in maternal physiology, progesterone reduces the contractility and excitability of uterine muscle and thereby helps to prevent pre-mature expulsion of the foetus from the uterus.

Androgens and Estrogens: Progesterone is converted by the foetal adrenal to adrenocortical steroids, as well as a variety of Q_{19} androgens such as dehydroepiandrosterone. An additional hydroxyl group is added to this weak androgen by the foetal liver, and the 16-hydroxylated compound is then passed back to the placenta where it is converted to *oestriol,* the main estrogen secreted by the human placenta. As the human placenta lacks the enzyme 16-hydroxylase, it is incapable of converting dehydroepiandrosterone directly to oestriol. The human placenta also synthesises and secretes smaller quantities of estradiol-17 b and oestrone from other C_{19} precursors such as androstenedione.

Like ovarian estrogens, placental estrogens promote the growth and development of the breast and the uterine muscle in pregnancy, in addition to inducing widespread changes in maternal metabolism which are reflected by alterations in proteins synthesis, carbohydrate metabolism, and adrenal and thyroid function.

Human Chorionic Gonadotrophin (HCG): The human placenta also secretes large quantities of HCG, a protein hormone. It is a glycoprotein of molecular weight around 30,000. The chemical structure and properties of HCG are very similar to those of the pituitary LH, although it has a higher carbohydrate content (8.5 per cent sialic acid). It is composed of at least two subunits one of which is common to FSH, LH, and TSH, accounting for the immunological similarity of these protein hormones. It was first demonstrated by Aschheim and Zondek in 1927 that pregnancy in the human female is characterised by the appearance in the urine of HCG which reaches its maximum concentration between the 60th and 80th days of pregnancy. This hormone has almost the same effects on the sexual organs as LH. Its most important function is to prevent the normal involution of the corpus luteum at the end of the female sexual cycle. Instead, it causes the corpus luteum to secrete even larger quantities of estrogens and progesterone for the next few months.

These sex hormones prevent menstruation and cause the endometrium to continue growing and to store large amounts of nutrients rather than being shed in the menstruum. Human chorionic gonadotropin also exerts an *interstitial cell-stimulating effect* on the testes, thus resulting in the production of testosterone by fetal tests. This small secretion of testosterone causes the fetus to grow male sex organs. At the end of pregnancy, the testosterone secreted by the fetal testes also causes the testes to descend into the serotum.

Human Chorionic Somatomammotrophin (HCS): Another protein hormone of the human placenta is HCS. It was formerly known as human placental lactogen (HPL). It is synthesised by the syncytiotrophoblast cells of the placental villi, and is secreted almost exclusively into the maternal circulation at a rate of about 1 g per day. It is similar immunologically and chemically to human growth hormone having a molecular weight of around 38,000.

It promotes development of the breast, and together with HCG is probably responsible for the maintenance of the corpus luteum of pregnancy. It also produces changes in carbohydrate and protein metabolism similar to those of the growth hormone. It causes decreased insulin sensitivity and decreased utilisation of glucose in the mother, thereby making more quantities of glucose available to the fetus for its growth. Furthermore, the hormone promotes release of free fatty acids from the fat stores of the mother, thus providing an alternative source of energy for her metabolism.

Pregnant Mare's Serum Gonadotropin (PMSG): This protein hormone is secreted by the endometrial cups of the pregnant uterus in horse and donkey. The molecular weight of PMSG is larger than that of any other gonadotrophin (about 70,000). The high molecular weight together with high carbohydrate content (49 per cent with 10.4 per cent sialic acid) impairs its clearance from the circulation and accounts for its long half-life and high concentration. It has properties of both the FSH and LH, but the former predominates. However, its physiological function is not fully known and its secretion is regulated by the genotype of the foetus.

Few more Endocrine Organs

Pineal Gland: The pineal is situated on a stalk between the two halves of the brain. In some vertebrates, for example, lampreys and tuataras, the pineal is a quite large structure. Here it is located directly under the skin of the head, and the skin at that point is transparent. In these animals the pineal is actually the third functional eye situated on the top of the head. The pineal gland plays an important regulatory role in sexual and reproductive function at least in some lower animals. The pineal gland is photosensitive and is controlled by the amount of light or "time pattern" of light seen by the eyes each day. The *melatonin,* secreted by the pineal gland, is believed to pass either by way of the blood or through the fluid of the third ventricle to the adenohypophysis to suppress gonadotropic hormone secretion. This occurs during the early winter months. Thereafter, the gonadotropic hormone secretion breaks through the inhibitory effect of the pineal gland and the gonads become functional.

In humans, tumours often occur in the region of the pineal gland. Such tumours are often associated with serious hypogonadal or hypergonadal function. Therefore, it appears that pineal gland plays atleast some role in controlling sexual drive and reproduction in humans.

Thymus Gland and its Role

It lies in the neck and the upper chest regions. In many mammals it persists throughout the life, but in man and some other mammals the organ gradually decreases in size during youth and disappears altogether at the time of puberty. The most important function of the thymus gland is that it pre-processes the *T-lymphocytes.* The T-lymphocytes, after their origin in the bone marrow, first migrate to the thymus gland. Here they divide rapidly and develop extreme diversity for reacting against different specific antigens. The different types of processed T-lymphocytes now leave the thymus and spread throughout the body in the lymphoid tissue. The thymus also makes certain that any T-lymphocytes leaving the thymus will not react against proteins or other antigens that are present in the body's own tissues. Most of the processing of the T-lymphocytes in the thymus occurs shortly before birth of a baby and for a few months after birth. However, removal of the thymus several months before birth can prevent development of all cell-mediated immunity.

Bibliography

Achaya, K.T. : *Indian Dairy Products,* Asia Publishing House, Bangalore, 1974.

Adam, H. : *Food Science,* John Wiley and Sons, New York, 1982.

Addis, Ababa : *Education and Training in Science and Technology in Asia,* UN Economic and Social Council, New York, 1989.

Arbuckle, S. : *Learning of Education,* Allyn and Bacon, Boston, 1965.

Bachman, W. : *Food Chemistry,* Maclaren and Sons Ltd., London, 1955.

Balsdon, J.P. : *Life and Leisure in Ancient Rome,* Bodley Head, London, 1966.

Banerjee, N. : *Indian Females in a Changing Industrial Scenario,* Indo-Dutch Series on Development Alternatives, Sage Publications, New Delhi, 1991.

Cain, M.L. : *Women and Technological Change in Development,* Westview Press, Colorado, 1981.

Carlton, E. : *Foundations of Technical Education,* Prentice Hall, New Jersey, 1964.

Carr, M. : *Appropriate Technology: Two Essays,* Intermediate Technology Development Group, London, 1982.

Crites, J.C. : *Vocational Psychology,* McGraw Hill, New York, 1969.

Donald, E. : *Learning of Technical Education,* Harper Row, New York, 1957.

Dumazedier, J. : *Towards a Society of Leisure,* Free Press, New York, 1967.

Duncan, W.A. : *Engendering School Learning: Science, Attitudes and Achievement among Girls and Boys in Botswana*, University of Stockholm, Stockholm, 1989.

Dyke, S.F. : *The Chemistry of Vitamins*, Interscience Publishers, New York, 1965.

Eley, G. : *Wild Fruits and Nuts*, E.P. Publishing Ltd., Yorkshire, 1976.

Franklin, U.M. : *Will Women Change Technology, or will Technology Change Women?*, The Canadian Research Institute for the Advancement of Women (CRIAW), London, 1985.

Gill, L.S. : *Science, Technology and Gender*, Polity Press in Association with the Open University, Cambridge, London, 1992.

Hanson, J.C. : *Vocational Guidance and Career Development*, MacMillan, New York, 1968.

Helen, Appleton : *Do it Herself: Women and Technical Innovation*, IT Publications, Mumbai, 1995.

Hoyt, K. : *Techniques of Education*, Science Research Associates, New York, 1959.

Hudson, F. : *Development in Food Proteins*, Applied Science Publisher, London, 1982.

James, A.R. : *Education : A System for the Seventies*, Charles E. Marrill Pub-lishing Co., Columbus, 1970.

Jones, A.J. : *Principles of Education in India*, McGraw Hill, New York, 1963.

Kak, Subhash C. : *Indian Physics: Outline of Early History*, ArXiv, London, 2003.

Khosbed, A.W. : *Education Movement in India*, NCERT, New Delhi, 1963.

Krause, M.V. and L.K. Mahann : *Food Nutrition and Diet Therapy*, W.B. Saunders Co., Philadelphia, 1979.

Krouse, H. : *Counselling Technique with Youth*, Charles and Mervil, Ohio, 1960.

Kumari, P. : *Role of Anganwadi Workers in Integrated Child Development Services Programme*, Discovery Publishing House, New Delhi, 1994.

Lagerstrom, B. : *Indian Women in Industry*, Zimbabwe House (Pvt) Ltd., New Delhi, 1984.

Maclean, M.S. : *Counselling of Technical Education*, McGraw Hill, New York, 1955.

Malcolm, S. : *Science and Technology for People*, Tamil Nader State Council for Science and Technology, Madras, 1985.

Marilyn, Carr : *Technology and Women: Conceptual and Empirical Issues*, Allen and Unwin, London, 2000.

Ndunda, M.M.K. : *Women's Resistance to Science Careers in India*, Queen's University, Kingston, 1990.

O'Kelly, Elizabeth : *Simple Technologies for Rural Women in Bangladesh and India*, Kuber Pub., Dhaka, 1978.

Oblsen, M.M. : *Technical Education*, Holt Rinehart and Winston, New York, 1977.

Pacey, A. : *Culture of Technology*, Cambridge, London, 1983.

Pinto, Diana V. : *Education : Theory, Research and Practice*, Rand McNally College Publishing Company, Chicago, 1978.

Potter, N.M. : *Food Science*, The AVI Publishing Co., Connecticut, 1978.

Rahman, A. : *Fatullah Shirazi, a Sixteenth-Century Indian Scientist*, MacMillan, London, 1968.

Rakowski, C.A. : *Women in Non-Traditional Industry: the Case of Steel in Ciudad Guayana*, Venezuela, London, 1985.

Samba, Murthy K.V. : *Science and Technology in Medieval India*, Pratap, Agra, 1982.

Sandler, J. : *The Tech and Tools Book: a Guide to Technologies Women are Using Worldwide*, International Women's Tribune Publications, New Delhi, 1986.

Sharma, P.V. : *History of Medicine in India*, Academic, New Delhi, 1992.

Singh, Harkishan : *Pharmaceutical Education*, Vallabh Prakashan, New Delhi, 1998.

Sponsors : *Proceedings of the Southeast Asian Seminar on Women and Science in Developing Countries*, Kovalevskaia Fund, New York, 1987.

Stadlmare, W.J. and. Cotterill, O.Y. : *Egg Science and Technology,* The AVI Publishing Co., Connecticut, 1973.

Stolte-Heiskanen, V. : *Science: Role of Changing India,* ISSC, New Delhi, 1991.

Subbarayappa, B.V. : *Science and Technological Exchanges between India and Soviet Central Asia,* Educational, New Delhi, 1985.

Theresa, Montini : *Science: Technology and Human Values,* Oxford, New York, 1993.

Tiwari, P.V. : *Kasyapa Samhita or Vrdhajivaka Tantra,* Visvabharati, Varanasi, 1997.

Tolber, E.L. : *Counselling for Career Development,* Houghton, Mifflin, 1974.

Uberoi, Chanchal : *India's Long History of Women in Science,* Physics Today, New Delhi, 1993.

Van, William N. : *Education in the 21st Century,* Houghton Mifflin Co., Boston, 2002.

Veken, M. Van Der : *Women: Technology and Development,* ATOL, Louvain, 1988.

Veldhnis, M.K. : *Food Science and Technology,* The AVI Publishing Co., Connecticut, 1977.

Vijaya Lakshmi, Sharma : *Science in Pre-Modern India,* D.K. Print World, New Delhi, 2001.

Virginia, Hazzard : *Village Technology and Women's Work in Eastern Africa,* UNICEF, Geneva, 1976.

Wardha, B. : *Science and Technology in India,* Centre of Science for Villages, Wardha, 1983.

Whitaker, J.R. : *Principles of Enzymology for Food Science,* Marcel Dekker, New York, 1977.

Winslow, A. : *Science: History,* Duke University Press, London, 1990.

Yukic, T.S. : *Fundamentals of Recreation,* Harper Row, London, 1963.

Zimmerman, J. : *Science in India: Today and Tomorrow,* Praeger Publishers, London, 1983.

Index

E

F

G

H

I

□□□